'This delightful introduction successfully fuses history, prehistory and earth science. It captures the imagination from its first page, and then takes the reader on a fun- and fact-filled world tour through the past.'

Professor Tim White, University of California at Berkeley, USA

'A fascinating guide to the measurement of time.'

Chemistry World

'… absorbing – will appeal to a wide audience, particularly those who got a kick out of Blink or Freakonomics.'

Publishers Weekly

'What I like best about the book: It's a scientist clearly explaining what he does for a living and why it is important, at a level that any literate person can understand. Not an easy accomplishment.'

scienceblogs.com/pharyngula

'Well researched and covers a lot of ground in a splendidly personal style. Highly recommended.'

Quaternary Australasia

D0043596

BONES, ROCKS AND STARS

The Science of When Things Happened

Chris Turney

Macmillan
London   New York   Melbourne   Hong Kong

First published in hardback 2006
First published in paperback 2008 by
Macmillan
Houndmills, Basingstoke, Hampshire RG21 6XS and
175 Fifth Avenue, New York, N.Y. 10010
Companies and representatives throughout the world

ISBN-13: 978–1–4039–8599–6 hardback
ISBN-10: 1–4039–8599–5 hardback
ISBN-13: 978–0–230–55194–7 paperback
ISBN-10: 0–230–55194–7 paperback

This book is printed on paper suitable for recycling and made from fully managed and sustained forest sources. Logging, pulping and manufacturing processes are expected to conform to the environmental regulations of the country of origin.

A catalogue record for this book is available from the British Library.

A catalog record for this book is available from the Library of Congress.

10   9   8   7   6   5   4   3   2   1
17 16 15 14 13 12 11 10 09 08

Printed and bound in China

To Annette, my ever-patient wife

I have measured out my life with coffee spoons
THOMAS STEARNS ELIOT   (1888–1965)

CONTENTS

LIST OF FIGURES AND TABLES

LIST OF PERMISSIONS AND FIGURE SOURCES

Figure 4.3 entitled 'Dating the Egyptian pyramids of the Fourth and Fifth Dynasties' came from, Spence, K. (2000) Ancient Egyptian chronology and the astronomical orientation of pyramids, *Nature*, **408**, 320–4.

The data used to plot part of the radiocarbon calibration curve used in Figure 5.1 'Using radiocarbon wiggles to date the Santorini eruption' came from Reimer, P.J., Baillie, M.G.L., Bard, E., Bayliss, A., Beck, J.W., Bertrand, C.J.H., Blackwell, P.G., Buck, C.E., Burr, G.S., Cutler, K.B., Damon, P.E., Edwards, R.L., Fairbanks, R.G., Friedrich, M., Guilderson, T.P., Hogg, A.G., Hughen, K.A. and Kromer, B. (2004) IntCal04 terrestrial radiocarbon age calibration, 0–26 cal kyr BP. *Radiocarbon*, **46**, 1029–58.

The data used to plot Figure 7.2 'Changing ice volume and solar radiation for the past 600,000 years' came from Berger, A. and Loutre, M.F. (1991) Insolation values for the climate of the last 10 million years. *Quaternary Science Reviews*, **10**, 297–318 and Imbrie, J., Shackleton, N.J., Pisias, N.G., Morley, J.J., Prell, W.L., Martinson, D.G., Hayes, J.D., MacIntyre, A. and Mix, A.C. (1984) The orbital theory of Pleistocene climate: support from a revised chronology of the marine $\delta^{18}O$ record. In: *Milankovitch and Climate*, Part 1, Ed. by A. Berger, Reidel, Hingham, Massachusetts, 269–305.

The data used to plot Figure 7.3 'Temperature changes in Greenland over the past 90,000 years' came from Blunier, T. and Brook, E.J. (2001) Timing of millennial-scale climate change in Antarctica and Greenland during the last glacial period. *Science*, **291**, 109–12.

Many thanks to Mike Baillie for permission to reproduce the illustration in Figure 6.1 entitled 'Oak ring patterns for trees growing during the 1628 BC event at Garry Bog, Northern Ireland'. This figure was modified from that published in Baillie, M. (2000) *Exodus to Arthur*, Batsford, London.

Every effort has been made to trace all the copyright holders but if any have been inadvertently overlooked the publishers will be pleased to make the necessary arrangements at the first opportunity.

ACKNOWLEDGEMENTS

In writing this book, I owe a great deal to the numerous texts listed under Further Reading. In addition, I am grateful to the many students, colleagues and friends I have had the pleasure of working with over the years. I would particularly like to thank the following individuals: Julian Andrews, Fachroel Aziz, Mike Baillie, Tim Barrows, Mike Benton, Michael Bird, Nick Branch, Charlotte Bryant, George Burton, John Chappell, Steve Clemens, Ed Cook, Alan Cooper, Joan Cowley, Margaret Currie, Siwan Davies, Charlie Dortch, Keith Fifield, Tim Flannery, Mike Gagan, Rainer Grün, Simon Haberle, Valerie Hall, Doug Harkness, Christine Hertler, Peter Hill, Doug Hobbs, Alan Hogg, Stephen Hoper, Mike Hulme, John Hunt, Sigfus Johnsen, the late Rhys Jones, Bob Kalin, Rob Kemp, Peter Kershaw, Dikdik Kosasih, Ollie Lavery, Finbar McCormick, Jim McDonald, Matt McGlone, Giff Miller, Neville Moar, Mike Morwood, Patrick Moss, Callum Murray, Colin Murray-Wallace, Jonathan Palmer, Jon Pilcher, Paula Reimer, Yan Rizal, Bert Roberts, Jim Rose, Ritchie Sims, Phil Shane, Mike Smith, Jørgen-Peder Steffenson, Chris Stringer, Djadjang Sukarna, Thomas Sutikna, Michelle Thompson, Chris Tomkins, Gert van den Bergh, Mike Walker, Stefan Wastegård and Janet Wilmshurst. A special thanks to John Lowe at Royal Holloway, University of London, for his years of inspired and level-headed professional advice without which I would not be where I am today. If I have forgotten anyone I am sorry.

I would also like to thank my editor Sara Abdulla at Macmillan for her guidance and patience in seeing this book through to the end.

Finally I would like to thank all my family, including my children Cara and Robert, and my parents Ian and Cathy. I am beholden to my darling and ever-patient wife, Annette, without whom this book would never have happened.

INTRODUCTION

Time present and time past
Are both perhaps present in time future,
And time future contained in time past

THOMAS STEARNS ELIOT (1888–1965)

Time is one of the greatest of all our obsessions. Why? In many ways, it's a complete paradox. After all, time has no physical basis. We can't feel or touch it. Yet there's almost a sense that we can see it. From as soon as we can remember, we become aware that 'time flies' and 'time is money'. We religiously follow the movement of the hands on a clock; we allow time to dictate our lives. And no matter how hard we try, most of us just don't have enough of it.

Unfortunately, we really can't ignore the unrelenting tick of the clock. Even a hermit living in the back of beyond isn't immune to its effects. Surviving the different seasons would force even the most zealous recluse to follow the demands of the clock. Regardless of whether it's a business meeting or the migration of a school of whales, our world runs on time. We simply can't avoid it.

How time is used has always been pretty controversial. The control of something we both love and hate has often been seen as a way of wielding power. When the world's clocks were set relative to Greenwich Mean Time in 1884, competing empires offered alternatives. When the modern Gregorian calendar was developed by the Roman Catholic Church in 1582, it was ignored by Protestant and other religious nations and resulted in organized chaos for several centuries.

Even what might seem to be a safe discussion on the age of our universe has got people into trouble. As recently as 2005, the singer Katie Melua had a top-5 hit single in the UK called

'Nine Million Bicycles'. One of the verses contained the rather innocent-sounding 'We are 12 billion light years from the edge. That's a guess. No one can ever say it's true'. We shall come back to the age of the universe later but for now let's just say the scientific community was incensed; this age was way off the mark. Interviews were had; a flurry of articles written. An alternative version was created, with the offending lyrics replaced by the less harmonious-sounding 'We are 13.7 billion light years from the edge of the observable universe. That's a good estimate with well-defined error bars. Scientists say it's true, but acknowledge that it may be refined'. Sometimes science and the arts just don't mix.

Fundamentally, we love to know how old things are. Every other day an article appears in a newspaper, on the web or on television, telling us that an archaeological or geological find has been discovered and it's 'x years old'. Big numbers are impressive, so ages regularly get top billing in the press. They grab the imagination. It almost seems that the further back in the past the better. But with this comes quite a bit of confusion. Although the example of Katie Melua and the age of the universe is a pretty small spat in the grand scheme of things, there is a difference of 1.7 billion years between the ages according to the lyrics and science. That's a heck of a long time.

During my scientific career, I've been fascinated by the past and communicating its importance but it does seem that there is an ever-widening gulf between enjoying the benefits of science and understanding it. Numbers are thrown about but it's not often clear how they were calculated. In many ways, this is true of countless branches of science. There's a danger that science is seen as too difficult, too boring. And it's not just the perception of time that's becoming an issue.

Perhaps the single greatest threat to twenty-first-century timekeeping is the pressure to teach 'creation science' in the school science classroom. This is the claim that the first book of the Bible, Genesis, is held to be the literal truth; with the most extreme form believing that God created the Earth in six days, just 6000 years ago. Fantastically, it just won't go away, despite all the evidence to the contrary. A recent NBC News poll in the US showed that 44% of adults believed in a literal biblical interpretation for the creation of the world. Clearly it's an idea that strikes a chord. That's fair enough. After all, it is a question of personal choice. Unfortunately, it's not often left to the individual; every now and again, some of its better funded believers gather their support and worryingly try to force their beliefs into school science classes. No one should claim that science has the answer to life, the universe and everything. But because of the way theories are constructed, tested and validated, the whole system is self-correcting.

The key word we hear with creationism is 'belief'. No matter how much science proves otherwise, some creationists still choose to *believe* the world is only 6000 years old. I might *believe* that the world is flat or that little green men live on Mars; should I get a teaching slot alongside electrostatics and gravity? I hope not.

We could argue: why does it matter? After all, the Western world has a good quality of life. Perhaps, but this is dangerously short-sighted. There are many challenges facing our world that urgently need to be sorted out. Massive extinction of the world's fauna and flora and extreme climatic change are just two examples where drastic action is needed by us all. If the Earth is only 6000 years old, many of the past catastrophes, which we will discuss later in the book, could not have happened. Our society is built on democracy but there are politics with time. If government, including educational policy, is hijacked by religious teaching, we're not giving ourselves a chance to learn from past calamities and face future challenges with any sort of confidence. Time gives us the framework to meet these challenges face to face, to manage them, to mollify and perhaps even prevent them happening.

These are exciting times in archaeology and geology. New techniques open ever-more windows into the past. Over the next 11 chapters, we'll take a look at how dating techniques have helped solve some of the most exciting mysteries of what has gone before: for us, our species and our planet.

Chapter 1

THE EVER-CHANGING CALENDAR

O aching time!
O moments big as years
JOHN KEATS (1795–1821)

The calendar we take for granted today has many a tale to tell. Spanning nearly 4000 years, it's had its fair share of ups and downs. Before the third millennium BC, the calendar hadn't really got going in a form we'd recognize today. The odd bone has been found, marked with enigmatic notches but no one can seem to agree whether these record the earliest means of timekeeping. Even if these marks did actually record days or nights, there doesn't seem to have been a widely accepted calendar that prehistoric people worked to. Most individuals probably just had to make do with a list of days numbered into the future from a fixed point of time. Anyone who didn't have a bone to hand would have had to make do with fingers and toes. That's no way to make any long-term plans. Fundamentally, our ancestors needed a calendar. But how to make one?

Two of the most important concepts needed for a calendar system are 'month' and 'year'. Now most people would agree that defining a 'month' as a full cycle of the different phases of the Moon sounds reasonable. The Babylonians, who inhabited what is roughly modern-day Iraq, certainly felt so and started using this system as far back as 3500 years ago. Each day began at evening, with the month starting on the first sighting of the crescent of a new Moon. This is a dependably regular 29.5 days and extremely tempting to use as the basis of a calendar. The first Babylonians did just that. Their calendar was made up of 12-lunar months of 29 and 30 days, and started during the northern hemisphere spring when the day and night are the same length: the vernal or spring equinox.

Using a variation of the Babylonian scheme, the Romans developed a 10-month calendar. This was supposedly started by one of their founding fathers, the warrior king Romulus, in 753 BC, the year of Rome's formation. In the Romans' scheme, the year began in March, with the months being named in a haphazard way. Even now we live with many of these original names, although some might seem a little odd for today's calendar – Aprilis, for raising pigs, Maius, for a provincial Italian goddess, Iunius, for the queen of the gods and, imaginatively: September, October, November and December for the seventh, eighth, ninth and tenth months of the year.

The problem both these civilizations realized, is that a calendar based purely on the changing phases of the Moon is not that accurate for tracking the seasons. To get over this, the Babylonians added the odd month now and again to keep things on course. The Romans had to be more drastic. They modified their 10-month calendar to include the months of Ianuarius and Februarius to try to make up the distance. But for the Romans, there still remained an alarming, ever-increasing difference between the seasons and the time of the year. The penny finally dropped that a 'pure' lunar calendar was no way to define a year.

An alternative way of defining a year is the length of time it takes the Earth to rotate around the Sun. One way to do this is to measure the time between two successive vernal equinoxes; the so-called tropical or solar year. Today, the tropical year is 365 days, 5 hours and approximately 49 minutes. This 'year' is a whole 11 days longer than one of 12-lunar months. After just 16 years, summer in a lunar-based calendar would be in the middle of the winter season. This was absolutely hopeless for long-term planning, especially in agriculture, which was a mainstay of the Roman economy.

In response, a group of Roman priests called the *pontifices* were tasked with keeping the calendar on track by adding days through the year. Although this sounds a great way of preventing any drift and keeping the system on track, there was another problem: the *pontifices* were notoriously corrupt. For years, no one beside the *pontifices* really understood the way the extra days were added and as result the system was ripe for abuse. Rather than including days in a predictable manner, the *pontifices* would frequently add or delay the introduction of days, and in some cases months, whenever it suited them; either for personal financial gain or to see their preferred candidates hold offices of power for as long as possible. Chaos frequently ensued.

By 190 BC, the Roman calendar was a full 117 days off, but somehow between 140 and 70 BC, the *pontifices* had managed to get the calendar back on track with the seasons. They soon lapsed again and by 46 BC, a 90-day difference had become the norm. Julius Caesar consulted astronomers about what to do. In 46 BC, the final 'Year of Confusion', Caesar made the changes necessary to get the system back on track. He added two temporary months and extended the length of the original 12 months to reach a makeshift total of 445 days. The following years would then run to 365 days, beginning in January. The jubilant public believed their lives had been extended by 90 days. More importantly, 45 BC was back in phase with the seasons.

Even with 365 days, this scheme did not fully capture a true year. Caesar argued that by adding an extra day every four years, the 'leap' year, he could correct for the missing six hours or so. This would keep the calendar on track with the seasons, or so Caesar believed. Shortly before Caesar's assassination in 44 BC, the Roman Senate was so impressed with the effectiveness of this long-overdue reform, it voted to rename one of the months Iulius, better known today as July, in his honour. Predictably, old habits die hard and after Caesar's assassination, there was a misunderstanding: the *pontifices* added the leap year once every three years. Only during Augustus Caesar's reign was this mistake corrected, by stopping the addition of leap years until the calendar was back on track after AD 8. For this and other political honours, the sixth month of the year was renamed Augustus, completing the full suite of month titles we use today.

This is not to say that there weren't other attempts to rename the months of the year. The Emperor Tiberius, in a moment of unusual discretion, overruled attempts by the Senate to rename September and October after himself and his mother. Commodus took quite a different tack and tried to have all the months altered to the other names of himself. Famously, December was changed to Amazonius after his obsession for the warriors of this name. Nero was a little more circumspect and only had Aprilis renamed Neronius to celebrate a failed assassination attempt. More recently, in the eighteenth century, the French revolutionaries had all the Roman names replaced by descriptions of the typical climate for each month. Thermidor, for instance, was the Hot Month. But this was totally hopeless for a country aspiring to an empire spanning different parts of the world. Unfortunately for those concerned, no one else felt quite the same about their stabs at calendrical immortality and any name changes after Augustus were soon dropped.

⧗

The Julian scheme is a reasonable first stab at a decent calendar, but at 365 days and 6 hours long, it does not track time as faithfully as might first appear. It gains 11 minutes on a real year. Over the course of one lifetime, an individual wouldn't notice the difference; it would take around 130 years before one extra day was gained. The problem was that over the long term, it did get noticed. By the mid-sixteenth century, the calendar had gained a total of 12 days against real time.

This shift had serious implications for the Christian calendar; most critically, which day to celebrate the most important religious event of the year – Easter? As Christianity spread across Europe and beyond, increasingly different biblical interpretations were being made as to when Easter should be celebrated. The Gospels were ambiguous as to when precisely the resurrection of Jesus Christ took place. Throw in the fact that the Gospels were recording the events using the Jewish, lunar-based calendar and confusion reigned. When should the celebration be made using the Julian calendar?

In AD 325, a meeting of Christian leaders at Nicea, in present-day Turkey, tried to reconcile these uncertainties. Finally a compromise was made. These early Church leaders decided to combine the phases of the Moon with the solar calendar devised by Julius Caesar. It was agreed that Easter would be the first Sunday after the first full Moon following the vernal equinox. The result has confused people ever since: the date of Easter varies each year and ranges from 'early' to 'late'. But the deed was done. Easter was forever linked to the timing of the vernal equinox.

In the mid-sixteenth century, a meeting of religious leaders at Trent in Switzerland finally agreed that the offset between the calendar and real time needed to be addressed urgently. They authorized Pope Gregory XIII to investigate. Gregory followed Caesar's lead and took advice from astronomers. In 1582, he proposed removing 10 days from October of that year. This set the vernal equinox to March 21, the recalculated date for this event when the agreement was made at Nicea, over a millennium earlier.

To make sure the calendar was self-correcting and the whole palaver never had to be repeated, the leap years were continued as before except at the end of each century: only one in four have an extra day added. As a result, 1600 was a leap year, but 1700, 1800 and 1900 lost the February 29 they would have had under the Julian calendar. The revised scheme only gains half a minute over a year and takes 2880 years before one day has to be added against real time. At last, the calendar truly matched real time. The Gregorian calendar had arrived.

Unfortunately for Gregory XIII, it was not a great time to establish a new calendar across Europe. The Reformation had started in 1517 when Martin Luther had pinned a list of complaints against the Church on the German cathedral of Wittenberg. Change had swept across Europe, which was now made up of a patchwork of Catholic and Protestant nations. The result was that when the changes were announced, most Catholic countries welcomed the Gregorian calendar and introduced it soon afterwards; Protestant countries were more wary. In Great Britain, Elizabeth I was enthusiastic but was stalled by Protestant clergy. Where the changes were made in Catholic Europe, it was often with comical results. In what is now Belgium, the correction was introduced on 21 December in 1582, resulting in the next day being 1 January 1583 and the entire population missing Christmas.

One of the fallouts of the change was that travelling only short distances between different European Christian states created significant problems. You could leave a Catholic country one day and arrive in a Protestant state before you had left. The offset between the calendars was magnified when going to Great Britain or its fledgling empire because of the difference in the date for the start of the year. Using a Gregorian calendar, the year began on January 1, but in Great Britain the traditional Julian year started on March 25. A traveller going from Continental Europe to Great Britain between January 1 and March 24 would, on paper, have gone back in time by a year.

Britain and her colonies only adopted the new calendar in September 1752; but by this time 11 rather than 10 days had to be removed from the calendar due to a century passing since its acceptance in Continental Europe. Many people were enraged at the loss of these 11 days. William Hogarth produced a print called *An Election Entertainment,* which has a banner demanding: 'Give us back our eleven days'. 'Time riots' were common, one of which in Bristol supposedly resulted in the deaths of several people.

This issue also had serious financial implications for those collecting tax and rents. During the first full year of the Gregorian system in 1753, bankers refused to pay the appropriate taxes until 11 days after the traditional date of March 25. The result: the British tax year started on 6 April and continues to do so; a relic of the great changes that took place over 250 years ago.

Other Christian countries and denominations remained surprisingly loyal to the Julian calendar. Although Sweden changed in 1753, just one year after Great Britain, many Eastern European countries did not change until the twentieth century: Greece only made the shift in 1924. The Eastern Orthodox Church continues with a variation of the Julian calendar, while nationally, Ethiopia continues to do the same, with no immediate plans to change.

Non-Christian countries and faiths felt even less urgency to adopt the Gregorian system. The Islamic religious calendar continues to be based on a lunar scheme and changes through real time: the New Year drifts from winter to summer over the course of 17 years. At a national level, Turkey only took on board the Gregorian dating system in 1926. China was later still, only accepting the scheme in 1949.

While it's all good fun to see how people have responded to the developments in the calendar over the years, we clearly haven't moved on that far. We're not immune to misunderstanding how it works. How many of us decided to celebrate the start of the new millennium on 2000 when there had never been a year zero? If nothing else, at least history does teach us we need to get the time right if we want to have a party.

Chapter 2

A HERO IN A DARK AGE

Lives of great men all remind us
We can make our lives sublime,
And, departing, leave behind us
Footprints on the sands of time

HENRY WADSWORTH LONGFELLOW
(1807–1882)

For a brief moment, dream of a world with a sword in a stone, knights in shining armour, a Round Table and a beautiful queen. Sound familiar? The popularity of the myths of King Arthur is curiously tenacious. Pre-Raphaelite painters were particularly obsessed, while *Star Wars* supposedly puts the story into the future. So strong is the image of Arthur it is easy to presume he was a medieval British bloke, albeit a chivalrous one. The problem is that the British leaders of the medieval period are all accounted for. There is literally no time left for Arthur to have existed. But what if we're wrong?

The key to whether there ever was a King Arthur lies in documents: books, letters and poems. But these are notoriously difficult to interpret. Although it's comforting to think of history as unbiased, it's not. Even today, we can read about world events and know we're only getting one particular point of view. Once we try going back in the past, this bias becomes even more difficult to detect. We no longer have a broad overview of different opinions, just a snapshot of views peppered through time.

Picture a humourless historian of AD 3000 discovering an ancient documentary called *The Holy Grail*, recorded by what appears to be an esteemed group of academics called Monty Python. Although the film was not made in the Arthurian period, our future historian might assume that there is some historical basis for the tale. It's not a huge leap of faith to then take the date of AD 932 from the beginning of the film as the date for King Arthur's existence. At the beginning of the documentary, Arthur introduces himself as King of the Britons and defeater of the Saxons. Using other sources, this would seem intriguing to our historian because at this time the German and Danish tribes that made up the Saxon race had conquered much of Britain. A Saxon king called Ethelstan was actually on the throne in England in AD 932. The point is: given enough time, common knowledge that seems obvious at the time can be lost and totally misinterpreted by future generations.

Some of the first popular stories of Arthur date from early medieval times and were written by an eclectic group of individuals. One of these was Geoffrey of Monmouth, a Norman–Breton cleric who rose to become a bishop towards the end of his life, whose *The History of the Kings of Britain* was 'published' in Latin in 1138. In stark contrast, Sir Thomas Malory, who wrote *Morte d'Arthur* (Death of Arthur) in 1470, was accused of murder, rape, extortion and robbery on more than one occasion. He only seems to have got around to writing *Morte d'Arthur* during one of his frequent sojourns in prison. Between them, we have the basis for most of the myths we enjoy today.

In these stories, Arthur reigns as 'King' or 'Emperor' of the Britons, inheriting the throne from his father, Uther Pendragon. Uther is said to have fallen in love with Ygerna, the wife of the Duke of Cornwall. While the Duke was fighting the King's troops, Uther uses Merlin's magic to successfully enter the castle of Tintagel and sleep with Ygerna. The result: Arthur. Depending on what you read, Arthur later pulls the sword from the stone or receives it from the Lady of Lake, and becomes king. A sort of Utopia then develops, with Arthur defeating the Saxons and creating a prosperous kingdom. He forms the knights of the Round Table which includes Sir Lancelot, Gawain and Galahad. Peace and prosperity reign. Arthur marries Guinevere and bases his court at Camelot.

It all seems so perfect, which is always a bad sign for the characters in a story. Things start to go terribly wrong: Lancelot and Guinevere have an affair. And as if that's not bad enough, a bastard son called Mordred turns up on the scene and raises an army to fight the King. To add to everyone's puzzlement, including no doubt Arthur's, there's quite a bit of confusion as to Mordred's name and his relationship to the King: he's also described as a nephew and called Medraut. The legions of Arthur and Mordred meet at Camlann and both leaders are mortally wounded. Arthur is taken over the sea to the Isle of Avalon to have his wounds tended. No more is heard of him but the myths maintain that he will return to save Britain in its hour of need; presumably better equipped than with a sword and a shield.

Monmouth's book is supposedly a history of the kings of Britain; the native Celts of England, Wales and southern Scotland. Monmouth taunts his readers by claiming in his introduction that he has translated a 'very ancient book written in the British language'. Yet, when you read it, you can't help but wonder if he has taken a little artistic licence with his writing. He seems to have scribbled down folklore, legends and poetry, put them together and somehow ended up with a book. Where Monmouth does refer to known historical characters, they appear in the wrong order or at the wrong events. He also makes a number of incredible claims: the first King of the Britons, Brutus, originally came from Troy; the Roman occupation of Britain never happened; three British kings sacked Rome; and Arthur invaded what was left of the Roman Empire. All great fun, but absolute rubbish.

Despite this, there are clues that parts of what Monmouth wrote might contain an element of the truth. He claims that Arthur was conceived in a Cornish castle called Tintagel.

Visited today, the twelfth-century Tintagel Castle ruins are an impressive sight, stuck out on a promontory into the Irish Sea and only accessed by a narrow path that falls away to the crashing sea below. The town is the closest you'll get to King Arthur Land, with car parks, cafés and shops all named after their famous association, and packed with hordes of tourists in the summer. Fortunately, even now, the narrow path does its job and keeps back many of the tourists from visiting the main site.

The Tintagel association gives us a good opportunity to see whether there is any truth behind Monmouth's claim. We don't actually have any copies of the first edition of Monmouth's book. The earliest version is the second edition of *The History of the Kings of Britain*, which was brought out in 1145. We don't know whether Tintagel was in the original rendition. Even though there's only a difference of seven years between versions, this could be significant if we want to take Monmouth seriously: Reginald, Earl of Cornwall, who built much of the castle after getting the land in 1141, was his half-brother. It is possible to believe that Tintagel was only included in the book after it had come into the family. Considering his track record it doesn't look good for Monmouth.

In spite of all this, excavations have taken place at the site over the past 50 years. These show that before the castle was built, the site was originally a Celtic monastery. Distinctive types of eastern Mediterranean pottery have been found, which show it was probably occupied sometime around the fifth or early sixth century AD. This is when Monmouth puts Arthur fighting the Saxons. In more support of Monmouth, excavations in 1998 by the University of Glasgow and English Heritage made a big media splash when a piece of slate was discovered with an inscription on it that included the name 'Artogonov' – dubbed 'Arthur's stone'.

If Monmouth was right that Arthur was fighting the Saxons, we should look at what was happening about this time in Britain and mainland Europe. For around three centuries, Britain had been part of the Roman Empire. The whole place seemed to have been pretty peaceful and prosperous. If there was a 'Made in Britain' stamp at this time, it would have been seen all over the Empire. The economy boomed. The beginning of the end seems to have taken place around AD 380 when the barbarians started getting serious: Scots (from Ireland), Picts (from Scotland) and Saxons, Angles and Jutes (from northern Germany and Denmark) all started to attack Britain at the same time. Fortunately, the 60,000-strong Roman legion forces withstood most of the attacks. By AD 395, however, the Roman Empire was having its own problems. After his death, Emperor Theodosius I had arranged for the empire to be split in two. He gave the eastern part to his son Arcadius (with the capital at Constantinople) and the western part to his other son Honorius (with the capital temporarily at Milan). By AD 406, the Visigoths from Germany had invaded Italy. In a desperate attempt to defend Rome, Honorius ordered most of the troops in Britain to be withdrawn. It was too little, too late: Alaric the Visigoth sacked Rome in AD 410.

What was left of Rome and the Roman Empire struggled on, severely weakened, and withdrew the last of its legions from Britain. Some attempt was made to keep a Roman presence, with the creation of a new post of *Comes Britanniarum*, Count of Britons, but this seems to have been merely an honorary role. The Count probably only had a small auxiliary force and couldn't be everywhere at once to deal with the mass attacks coming from almost every direction. By AD 418 the Empire gave up on Britain: it was declared independent and told to get on with looking after itself. Control went back to the ancient Celtic tribal chiefs. The Empire had enough on its plate: Rome was sacked again in AD 455 by another German tribe, the Vandals. The remains of the Western Roman Empire effectively collapsed after AD 476. It was a dark age and must have seemed like the end of the world to many.

Against this backdrop of chaos, there was a surprisingly large number of people writing. Not all of them seemed to have been that worried about what the actual year was, probably because they were more concerned with whether they were going to get a sword in their ribcage by lunchtime. But this poor reference to dates was also a habit among some later writers. Monmouth only mentions two events that can date Arthur, while Malory gives just AD 487 for the start of the fantastic quest for the Holy Grail. Can we sift through the rubbish in the early writings to work out what was going on?

Today, we take it for granted that all dates are given relative to the birth of Jesus Christ. This was not the case before the collapse of the Western Roman Empire. A Scythian monk called Dionysius Exiguus, known as Dennis the Little, only came up with the method we use today in the early sixth century AD. Dennis was not that interested in how to record years; his main concern was calculating when Easter should be celebrated. The Church was constantly getting in a tangle over it. Because the AD 325 Council of Nicea had agreed to link Easter with the Moon and the vernal equinox, hardly anyone knew how to make the calculations. To make matters worse, the early Church had started to show signs of disagreement that would eventually result in the East–West schism: they used different dates for the vernal equinox. Most of the time it didn't make a blind bit of difference but every now and again Easter in the East and West would be a week apart. Not good for the unity of the Church.

In AD 525, Dennis was told by the Church of Rome to calculate the date of Easter. Using calculations from Alexandria and a date for the vernal equinox of March 21, he published a table of Easter dates that agreed with the Eastern Church and then extended them, bringing at least a small measure of unity. But what was the best way to report the year?

Before Dennis, you could date a year almost any way you wanted. The Greek historian Timaeos introduced the concept of dating time by the number of Olympics; *Olympiad* in Greek meaning chronology. Another common dating scheme in the Christian world was to date the number of years since the death of Jesus Christ – the Passion – which would now be written as AD 28. When telling the masses when to celebrate Easter, the Church often used the number of years since the Roman Emperor Diocletian came to the throne, something we now know as AD 284. Dennis was not impressed; Diocletian was a well-known persecutor of the early Christian Church. He felt it was far better to date Easter relative to the birth of Jesus Christ. Later the terms 'BC' – before Christ – and 'AD' – Anno Domini, 'in the year of our Lord' – were introduced. Slowly the method spread to the fringes of Europe. Yet even in the fifteenth century, Malory gave the date for the start of the quest for the Holy Grail relative to the Passion.

The bottom line is that any document reporting an event before AD 525, or even sometime afterwards, has to be treated with extreme suspicion. Unfortunately, Dennis made a mistake. He calculated the birth of Jesus Christ as 25 December 1 BC, so that 1 AD fell as the first year of his life. Using early records, Dennis had seen that Christ was born in the 28th year of Augustus Caesar's reign. What he did not realize was that Augustus had been known as Octavian for the first four years of his leadership. Octavian had effectively led the Roman Empire from 31 BC, but only officially became emperor in 27 BC, when he changed his name to Augustus. Independent of this, we now know that King Herod died in 4 BC. Christ must have been born in 4 BC.

⌛

To really get a good fix on a 'historical' event, it has to be cross-checked against another source. For example, Monmouth only gives one actual date for the time of Arthur, his death in AD 542. But he also states three times that Arthur was in Gaul, present-day France, when Leo was emperor. We know Leo I reigned over the Eastern Roman Empire in Constantinople between AD 457 and 474. Confused? Monmouth certainly was.

Gaul at this time was in chaos, and formed a major setting for the final death throes of the Western Roman Empire. Although it was technically Roman, large areas had been invaded by barbarian hordes. Euric, King of the Visigoths had conquered Spain at the time and was threatening Gaul. Trying to prevent this, Leo I appointed the Greek Anthemius as Western Emperor in Rome to form an alliance with British forces to stop Euric's advance. Other documents confirm this actually happened.

It's at this time that we start to hear of a whole host of weird and wonderful names, some often spelt several different ways. I'll try and keep these to a minimum but we'll have to include some because they're pivotal to the story. The first of these was a leader called Riothamus, a 'King of the Britons', who formed the British part of the alliance to stop Euric.

To confuse matters, we now know that Riothamus is not actually a name but a title meaning 'Supreme King'. A letter written by Sidonius Apollinaris, the Bishop of Clemont-Ferrand in Gaul, was addressed to Riothamus around AD 470, placing this character at around the same time as Arthur. What happened was transcribed by Jordanes the Goth in his *Gothic History*:

> Now Euric, king of the Visigoths, perceived the frequent change of Roman Emperors and strove to hold Gaul by his own right. The Emperor Anthemius heard of it and asked the Brittones for aid. Their King Riotimus [Riothamus] came with twelve thousand men into the state of the Bituriges by the way of Ocean, and was received as he disembarked from his ships. Euric, king of the Visigoths, came against them with an innumerable army, and after

a long fight he routed Riotimus, king of the Brittones, before the Romans could join him. So when he had lost a great part of his army, he fled with all the men he could gather together, and came to the Burgundians, a neighbouring tribe then allied to the Romans. But Euric, king of the Visigoths, seized the Gallic city of Arverna; for the Emperor Anthemius was now dead.

Here the similarity to the Arthur of legend becomes very strong. A deputy-ruler later betrays Riothamus; Riothamus follows a line of retreat to Avallon in Burgundy; he then promptly vanishes without trace. Could Riothamus be King Arthur around AD 470?

Let's test the idea against other writers of the time. One source is the *Anglo-Saxon Chronicles*, compiled under the reign of Alfred the Great, AD 871 to 899. The *Chronicles* are based on a number of early west Saxon monastic records for the Arthurian period and are at best a faithful copy of the original texts. Of importance to us is the timing of the arrival of the Saxons in Britain, which was known as the *Adventus Saxonum* (see Table 2.1):

AD 449. In their days Hengest and Horsa, invited by Wurtgern [Vortigern], king of the Britons to his assistance, landed in Britain in a place that is called Ipwinesfleet [Ipswich]; first against them. The king directed them to fight against the Picts; and they did so; and obtained the victory wheresoever they came. They then sent to the Angles, and desired them to send more assistance. They described the worthlessness of the Britons, and the richness of the land. They then sent them greater support. Then came the men from three powers of Germany; the Old Saxons, the Angles, and the Jutes.

AD 455. This year Hengest and Horsa fought with Wurtgern the king on the spot that is called Aylesford. His brother Horsa being there slain, Hengest afterwards took to the kingdom with his son Esc.

Table 2.1 Key sources, events and dates for the Arthurian period

Key 'Arthurian' events	Key 'Arthurian' sources					
	Gildas *The Ruin of Britain*	Nennius *History of the Britons*	*Welsh Annals*	Bede *Ecclestiastical History of the English People*	*Anglo-Saxon Chronicles*	Geoffrey of Monmouth **The History** *of the Kings of Britain*
Vortigern's 'reign' begins	·	(AD 429 or AD 445–446)	AD 397			
Arrival of Saxons in Britain			AD 400	AD 449	AD 449	
Saxon uprising					AD 455	
Ambrosius Aurelianus' 'reign' begins		(AD 441 or AD 458)				
Arthur's 'reign' begins						(Between AD 457 and 474)
Battle of Mount Badon	(AD 493 or AD 501)		(AD 490 or AD 518)	AD 493		
Arthur at Camlann			(AD 511 or AD 539)			AD 542

Note: Dates in brackets are where links have been made to different historical sources

No Riothamus or Arthur is mentioned in either of these entries but this would not be expected in an enemy source recording events 20 years before AD 470. Instead, a different leader makes an appearance: Vortigern, which means 'foremost prince'.

Even today British school children are taught about Vortigern: he was the misguided fool who invited the Saxons to his country, which led to its downfall. We know he definitely existed because he also pops up in the *Annales Cambriae*, the

Welsh Annals. The copy we have today comes from around the early twelfth century, but the entries themselves appear largely unaltered from when they were first written:

> Vortigern held rule in Britain in the consulship of Theodosius and Valentinian. And in the fourth year of his reign the Saxons came to Britain in the consulship of Felix and Taurus, in the 400th year from the incarnation of Our Lord Jesus Christ.

When the Roman Empire divided in AD 395, both emperors could elect a right-hand man called a consul, who held the post for one year. This fact is quite useful for us when comparing texts and dating. So, if the *Welsh Annals* are to be believed, Vortigern was living around 50 years earlier than the *Chronicles* claim.

We know from other sources that the consulship of Felix and Taurus began in AD 428 and not AD 400. This difference of 28 years shows a common mistake. The original entry must have been made relative to the death of Christ and not his birth, as shown in the *Welsh Annals*. Even so, it seems unlikely that one Vortigern could have led all the separate British tribes for 30 years as suggested. Is it possible that Vortigern might actually be two individuals with the same title?

In the ninth century AD, the Welsh monk Nennius 'heaped' together what he could find from across Britain into the *Historia Brittonum*, the *History of the Britons*. Thankfully, Nennius doesn't seem to have tried to do anything with what he found. Instead, what we're left with are tantalizing scraps of different events. Nennius actually lists two versions of Vortigern's death. One story involves a visit by St Germanus of Auxerre who arrived in Britain and then duly burnt Vortigern to death in the leader's fortress. The second story ends differently for Vortigern: after inviting the Saxons, 'he wandered from place to place until at last his heart broke, and he died without honour'.

So, confusingly, it looks like there were two leaders with the same title, Vortigern: one probably becoming 'overlord' in AD 425 and dying during St Germanus's known visit in AD 445–46; the second dying of grief when his policy of using Saxon mercenaries had clearly failed (Table 2.2).

Table 2.2 Best-guess dates of key events for the Arthurian period

Key events	Best-guess dates
Vortigern 1 leadership commences	AD 425
Vortigern 2 leadership commences	AD 445–446
Arrival of Saxons in Britain	AD 449
Saxon uprising	AD 455
Ambrosius Aurelianus's leadership commences	AD 458
Arthur's leadership commences	After AD 470
Battle of Mount Badon	AD 490
Arthur dies at Camlann	AD 511

This fits in with another nugget of information from Nennius. A leadership challenge took place 'from the [beginning of the?] reign of Vortigern to the quarrel between Vitalinus and Ambrosius are twelve years'. If this is right, the second Vortigern would have ended his leadership 12 years after AD 445–46, that is, sometime around AD 458; three years after being defeated by Hengest and Horsa, when the policy of using Saxon mercenaries had clearly failed. Either way, neither of the Vortigerns could have been the hero known as Arthur.

So how can we find Arthur? Thankfully, other accounts start to shed some light on our quest. During the sixth century AD, one of the most depressed monks Britain ever produced was writing. Gildas wrote the closest thing to a contemporary account of this period in Britain, *De Excidio Britannia* (The Ruin of Britain), but this was not a history, nor a celebration of his country or philosophy. No, this was a long tirade against the British leaders of his time. Gildas seemed to need to complain about almost everything, including the loss of Roman life and the poor leadership among the Britons. He doesn't say when he is writing but he refers to one leader that we know died in a national plague in AD 549. This would place Gildas writing a few years earlier, let's say around AD 545, but this is a bit of a guess on our part.

According to Gildas, some time after the Romans had left Britain, the Britons pleaded for help:

> The miserable remnants sent off a letter again, this time to the Roman commander Agitus in the following terms: 'To Agitus [Aëtius] thrice consul, the groans of the British'. Further on came the complaint: 'The barbarians push us back to the sea and the sea pushes us back to the barbarians, between these two kinds of death, we are either drowned or slaughtered'. But they got no help in return.

Aëtius was one of the last great personalities of the Roman Empire, defeating Atilla the Hun in AD 451 during the last of his three consulships, which he held in Gaul. He was the first person to hold three consulships for over 300 years, and independent sources indicate that this period ran from AD 446 to 453.

Importantly, Bede's *Historia Ecclesiastica*, the *Ecclestiastical History of the English People*, also reports this event. Written around AD 731, it gives a Christian history of Britain using the Anno Domini system for the first time. The dates for Aëtius's third consulship must mean that the plea from the Britons was sent during the reign of the second Vortigern.

So where are we at? During the late AD 440s, Britain was facing repeated attacks from Picts and Scots. What was left of the Western Roman Empire was fighting a hopeless rearguard action on mainland Europe against barbarian hordes. Officially, Britain had been independent since AD 418 and Aëtius, the Roman commander of Gaul, couldn't or wouldn't help. The Britons, under the second Vortigern, resorted to what they had done before: they invited Saxon mercenaries to help them combat the marauding Picts and Scots. Unfortunately for the

Britons, the Saxons revolted this time and the British leadership split, uncertain as to how best deal with the onslaught.

It is around this time that Gildas describes a leader called Ambrosius Aurelianus, who apparently rallied the Britons against the Saxons:

> Their leader was Ambrosius Aurelianus, a gentleman who, perhaps alone of the Romans, had survived the shock of this notable storm: certainly his parents, who had worn purple, were slain in it.

This is a rare moment in Gildas's writing. He admired this leader. The name and the fact that his parents 'had worn purple' indicates Aurelianus was of Roman descent; he appears to have led a revival of sorts that seems to have lifted some of Gildas's gloom. Bede also mentions him, although his text is identical to that of Gildas, suggesting he was paraphrasing the depressed British monk.

We now have some idea of when the two Vortigerns reigned. We also know from Nennius that there was a battle for the leadership of the Britons, apparently won by Ambrosius Aurelianus. This suggests that Aurelianus's 'reign' probably started around AD 458. This is close to when Riothamus was in Europe. According to Bede:

> Under his leadership [Aurelianus] the Britons took up arms, challenged their conquerors to battle, and with God's help inflicted a defeat on them. Thenceforward victory swung first to one side and then to the other, until the Battle of Badon.

Bede implies that Ambrosius Aurelianus led the Britons to victory at a major battle called Badon, although whether he was the leader is ambiguous. Other sources suggest a different scenario. Nennius identifies Badon as one of 12 battles and links them all to Arthur:

At that time, the Saxons grew strong by virtue of their large number and increased in power in Britain. Hengist having died, however, his son Octha crossed from the northern part of Britain to the kingdom of Kent and from him are descended the kings of Kent. Then Arthur along with the kings of Britain fought against them in those days, but Arthur himself was the military commander [dux bellorum] ... The twelfth battle was on Mount Badon in which there fell in one day 960 men from one charge by Arthur; and no one struck them down except Arthur himself, and in all the wars he emerged as victor.

Badon was the Battle of Britain of its day and seems to have been a major turning point after a succession of indecisive encounters. The battle site has never been found but was probably somewhere on the hills surrounding Bath. It was strategically placed against a Saxon advance from the east. The Britons had to win. A loss would have resulted in the Saxons driving a fatal wedge between the remaining British kingdoms in the west. Instead, the opposite happened, apparently thanks to Arthur. The Saxons were decisively defeated and they fell back. The archaeological record indirectly supports this. There is an almost complete absence of Saxon pottery over a 50-year period in the Thames valley during the sixth century AD. Rudolph of Fulda also records the rare occurrence of Saxons returning to the mouth of the Elbe from Britain sometime around AD 530. All of this points to an overwhelming British victory sometime around the start of the sixth century AD.

Was 'Arthur' a name or a title? There are at least two examples of Roman soldiers who served in Britain with the name 'Artorius'; one from the second century AD who formed the basis of the character 'King Arthur' in the 2004 Hollywood movie. Either could have left descendents to which Arthur was related. Alternatively, Arthur may have been a title of sorts. In Welsh, the word for 'bear' is *arth*, in Latin, *ursus*. Arthur, then, might be a blend of the two synonyms: Arthursus. Several Britons are known to have held Roman and Celtic versions of the same name. Arthur could have done the same, forming a title to please Britons of Celtic and Roman persuasions.

ፘ

Parallel to all the chaotic changes taking place in politics and war, yet another method was being used for dating: the number of years into an Easter cycle. The Easter cycle is the 532 years it takes for the celebration to take place on the same day of the month with the same phase of the Moon. Because of the complexity in calculating when to celebrate Easter, the tables produced by Dennis the Little and others were sent to all the centres of learning and worship so that everyone was singing from the same hymn sheet. All the clergy members had to do was remember the cycle year and they could read off what date to celebrate. But the tables soon assumed a historical significance: clergy would often scrawl events of the year against the entry.

Now if 'Arthur' was a title, maybe Ambrosius Aurelianus was him from around AD 458? To answer this, we can turn back to the *Welsh Annals*, but here the relevant sections are given against part of an Easter cycle:

Year 72: The Battle of Badon, in which Arthur carried the cross of our Lord Jesus Christ on his shoulders for three days and three nights, and the Britons were victorius.

Year 93: The strife of Camlann in which Arthur and Medraut perished, and there was plague in Britain and in Ireland.

Although these ages appear to be 'floating', all is not lost. Several other events are also recorded that can be fixed in time.

In Year 9: 'Easter is changed on the Lord's Day by Pope Leo, Bishop of Rome'. This event was one of the frequent spats between the Eastern and Western Churches as to when to celebrate Easter, before Dennis the Little published his tables. This change by Leo is known to have taken place in AD 455. Back calculating, we get the date AD 446 for Year 1. From this, the dates AD 518 and AD 539 can be calculated for Years 72 and 93. If Ambrosius Aurelianus started his leadership in AD 458, the dates in the *Welsh Annals* show he cannot have been Arthur.

Gildas does not mention Arthur by name but we can use his text for the legendary king. Gildas mentions 'the siege of Mount Badon', but his style of Latin is notoriously difficult to read and the relevant passage could have two interpretations: the battle could have taken place 44 years before the time of writing or 44 years after the Saxons arrived. If Gildas meant 44 years before he was writing, it suggests Badon took place around AD 501, significantly different to the *Welsh Annals* date of AD 518. Bede is more explicit than Gildas. He mentions the 44 years but states it was from the *Adventus Saxonum*. This is dated at AD 449, putting Badon at around AD 493.

The Badon dates given by Bede and implied by Gildas are quite a bit earlier than that suggested by the *Welsh Annals*. If the dates relevant to Arthur in this part of the *Welsh Annals* were incorrectly copied and refer to the years since the Passion, we can take 28 from the Years 72 and 93: Badon would now be AD 490 and the death of Arthur at Camlann, AD 511. This new date for Badon is very close to Bede's AD 493 and is probably the closest we can hope to get to for this time in Britain's history (Table 2.2).

Although we can't say with certainty who Arthur was, it seems likely that there was a leader of the Britons sometime around the end of the fifth century and the start of the sixth century. This 'Arthur' appeared to have led a series of Roman–Celtic victories over the invading Saxon forces. If so, the status of Arthur as a final figure of victory must have been of supreme importance to the Britons. Over time, the stories were embellished until they turned to legend, particularly once the British defences collapsed soon after Arthur's death. By AD 580, Durham, Bath, Cirencester and Gloucester had fallen, resulting in the long-term control of most of Britain by the Saxons. The result: the Britons themselves became isolated in what the Saxons referred to as the land of the foreigner, 'Weala', now known as Wales. The effect of a British victory at Badon could only last so long.

Up to the end of the sixth century, there is no record of anyone in Britain being called Arthur, but soon after, at least six Britons are known to be so named. It was suddenly fashionable to have the same name as the famous leader, much as today when people call their children after actors or pop stars. The seventh century poem 'Y Gododdin' by the bard Aneirin celebrates a British hero who fought at the Battle of Catterick around AD 600: 'He fed black ravens on the rampart of a fortress, though he was no Arthur.' The seeds were planted for what was to become the legend of King Arthur.

Chapter 3

THE FORGED CLOTH OF TURIN

Antiquities are history defaced,
Or some remnants of history
which have casually escaped the shipwreck of time
FRANCIS BACON (1561–1626)

The Turin Shroud is one of the most instantly recognizable religious relics in the world. A linen cloth 4.4 m by 1.1 m, the Shroud bears the front and rear image of a bearded man who appears to have been crucified and then wrapped in a cloth before burial. With the strong parallels to the death of Jesus Christ, the direct dating of the Shroud was often believed to be the definitive test of its authenticity. Yet before the results of the scientific analysis hit the headlines in 1989, the Shroud had had a colourful story.

The cloth first appeared in historical records sometime around 1350, although the date seems to vary depending on which historical source is interrogated. It appears to have been originally owned by a knight called Geoffrey de Charney from Lirey, in eastern France. How he came to gain possession of the cloth is unknown. Little else is known about him except that he was the author of the only book on chivalry at the time. De Charney died during the Hundred Years War at the Battle of Poitiers in 1356, leaving a widow and infant son. Searching through de Charney's belongings, his widow discovered the cloth and had it placed in the local church. By 1357, the first pilgrims were known to be visiting Lirey to see the cloth as the Shroud of Christ, bringing some much-needed money to the de Charney family and the local area.

Even at this time, the Shroud was not without its controversy. Several times it was declared a fake, including by two local bishops. One wrote in a letter that he knew who had done the forgery, though the suspect was not named. His successor wrote to the Avignon pope to request it be removed from public display because of this accusation. Despite all the fuss, the Shroud continued to be shown to pilgrims and remained in the family until it was sold to Duke Louis I of Savoy in 1453. Louis had it moved to his base in Chambéry, in southeastern France.

In 1532, a major fire scorched the linen as it lay in a silver chest in a chapel. Fortunately, the cloth was saved but not before part of the chest had melted and dripped silver onto parts of the image; scorch marks can still be seen on the Shroud today. The chest was doused in water before any further damage could take place.

In 1578, the House of Savoy moved to Turin in Italy, and the Shroud has been there ever since. It remained largely forgotten until 1898, when an Italian photographer called Secondo Pia took some pictures of the Shroud. To his surprise, Pia found that the image was a photographic 'negative' of a crucified man and revealed far greater details than had previously been seen. Suddenly there was renewed interest in this relic: how could it have been formed? This curiosity has endured to today.

In 1983, the final change of ownership took place: King Umberto II, who was a member of the House of Savoy, willed the Shroud to the Vatican under the custody of the Archbishop of Turin. It is now permanently stored behind the main altar of the Cathedral of John the Baptist in Turin. This much is certain.

There are several early stories of cloths bearing images of Jesus Christ. One legend has it that Christ's burial cloth was taken to King Abgar V of Edessa in southeastern Turkey after the resurrection. What happened to this particular cloth is unclear but in the first half of the sixth century, a similar item was discovered, supposedly in the walls of Edessa, sometime during either 525 or 544. Unsurprisingly, the cloth was instantly recognized as a relic and a church was built to house it. Here the cloth lay for several centuries until the Eastern Roman Emperor Romanus I apparently sent an army to Edessa in 944 and took the cloth back to Constantinople. Then the cloth simply disappeared from the history books. Needless to say, some have suggested that the Edessa image and the Turin Shroud are one and the same and that de Charney must have collected his cloth while visiting Constantinople as a knight; but this is pure speculation.

Over the years, people have become fascinated by the origin and age of the Turin Shroud. Some researchers have pointed out that the Shroud has a similar image to the decorated funeral sheet of Christ of Limutin Ure, made sometime between 1282 and 1321 and housed in the Museum of Church Art in Belgrade. Other funeral sheets with similar images also date back to the eleventh century. Could one of these sheets have been copied to make the Turin Shroud? An excellent test would be to date the cloth itself. An age of around 2000 years would support the idea that the Turin Shroud was what it was claimed to be. Radiocarbon was the ideal candidate for the dating.

$$\boxtimes$$

Radiocarbon dating is a way to work out the age of any material that contains carbon and was formed up to 60,000 years ago. It's probably one of the best known of all the dating methods and has revolutionized our understanding of the past.

Before we embark on how radiocarbon dating was applied to the Turin Shroud, we need a quick recap of some basic principles of radioactive decay. Somewhat like the solar system, atoms comprise a nucleus made up of protons and neutrons, orbited by electrons. Importantly, elements are distinguished by the number of their protons. The simplest and lightest, hydrogen, has one. To describe an element in shorthand, we use one or two letters, for example 'H' for hydrogen. The number of protons and neutrons are added together, known as the mass number, and placed to the upper left of the element's letter(s). The simplest form of hydrogen is the odd one out of all the elements in the periodic table: it has no neutrons and just one proton, so is written as ^1H.

In most cases, a balance exists between the number of protons, neutrons and electrons, making the atom stable. Although elements are characterized by the number of their protons, there are variations on a theme: atoms with different numbers of neutrons are called isotopes. In these cases, the letter stays the same but the mass number can change. So, using the example of hydrogen, there is a stable version called deuterium, written as ^2H, which has one proton and one neutron. But, as the number of neutrons increases, there is a greater chance that the combination will become unstable. When this point is reached, the atom will disintegrate, giving off one of a number of different particles or energy forms, in its quest to reach a more stable form. In the example of hydrogen, tritium, which is written in shorthand as ^3H, has a combination of one proton and two neutrons and is thoroughly unstable: it must break down.

Our understanding of radioactivity has a pretty short history. It was only in 1895 that German scientist Wilhelm Röntgen observed X-rays as a new source of energy when they caused a specially coated paper to glow. In 1896, the French scientist Henri Becquerel reported similar rays originating from uranium salts. By 1898, the Polish and French scientific partnership of Marie and Pierre Curie described similar effects from thorium and coined the term 'radioactivity'. Looking at the radioactivity of another mineral, pitchblende, the Curies found it gave off more energy than pure uranium, suggesting to them that there were other radioactive elements present. Amazingly, the couple sifted through literally tonnes of waste pitchblende, which had been used to extract uranium but was still highly radioactive. By 1902, the Curies had managed to isolate two new radioactive elements that they called polonium and radium. Suddenly, radioactivity seemed to be everywhere.

Marie and Pierre shared the Nobel Prize for Physics in 1903 with Becquerel. Pierre Curie died shortly afterwards in 1906, having slipped in front of a horse-drawn wagon as a result of a dizzy spell, most probably brought on by years of radiation exposure. Marie Curie later got a second Nobel Prize for Chemistry in 1911 for her work on radium and lived on until 1934, aged 67 years. She eventually died of leukaemia as a result of radiation sickness. Her laboratory notebooks are still so radioactive that they are kept in a lead-lined safe. The Curies' discoveries laid the foundation for relativity, atomic and quantum physics and certainly revolutionized the way we pinpoint the past.

Radiocarbon dating builds on this and exploits the changes in the amount of the radioactive isotope of carbon over time. The two most common forms of carbon, ^{12}C and ^{13}C, make up virtually all types of modern carbon and are stable – ^{12}C is the simplest form and is made up of 6 protons and 6 neutrons; ^{13}C is slightly heavier because it has one more neutron. The version we're concerned with here is the radioactive form, ^{14}C. Commonly known as radiocarbon, it has the unstable combination of 6 protons (defining it as carbon) and 8 neutrons. Radiocarbon is miniscule and forms just one trillionth of all modern carbon. This is the equivalent of one drop of water in an Olympic-size swimming pool.

We'll look at some of the greats who pioneered the use of radioactivity to date the past later in the book (Chapter 11), but in the case of radiocarbon, we'll fast forward to the mid-1940s. It was at this time that pioneering American chemist

Willard Libby suggested that the minute amounts of radio-carbon came from the upper part of the atmosphere. Libby put forward the idea that when high-energy particles that formed deep in space – called cosmic rays – reached our planet, they interacted with nitrogen gas in the atmosphere to form radio-carbon. Libby argued that the newly formed radiocarbon was rapidly converted to carbon dioxide, CO_2, and then taken up by plants during photosynthesis. The result is that when an animal grazes and/or is eaten by another, radiocarbon atoms are taken up through the food chain. Everything alive should therefore have the same radiocarbon concentration as the atmosphere. But once the individual dies, some of the ^{14}C atoms begin to disintegrate and give off an electron to reform nitrogen (Figure 3.1). Libby argued that if the original radio-carbon content was known, it would be possible to measure the remaining ^{14}C in a sample to back-calculate its age. The principle is the same as inferring how much time has passed by measuring the sand left in the top of an egg timer.

By the end of the 1940s, Libby and his team had shown that the radiocarbon content of the air was the same around the world and that ^{14}C could be used to date anything organic. Soon they were making the first independent age estimates by measuring the amount of radiocarbon left in samples. Radioactive dating had arrived.

A crucial principle of all this is the rate at which an unstable atom breaks down: its half-life. Unlike living things that have an increasing chance of dying with age, radioactive isotopes can die at any moment. It's just a matter of proba-bility. The half-life is the time it takes for an original quantity of isotope to halve. This varies depending on what the isotope is; the more unstable the combination of protons and neutrons, the shorter the half-life. It sounds a bit abstract but let's take an extreme example to illustrate the principle. Imagine a laboratory where a scientist has a 1-kg sample of a radioactive isotope known to have a half-life of just 5

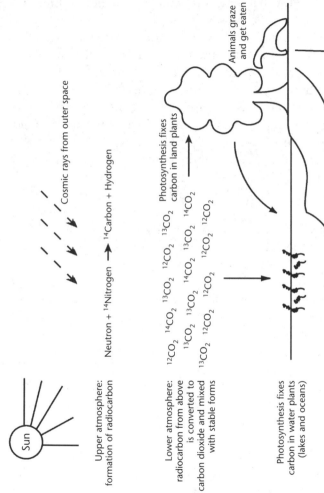

Figure 3.1 Radiocarbon formation and movement in the environment

minutes. During the first 5 minutes, the sample would start to disintegrate in front of her eyes: there would only be 500 grams left. A further 5 minutes on, only 250 grams would be left. After a further 5 minutes on, 125 grams. With each half-life, the sample literally halves in quantity. This would continue until, after around 10 sets of 5 minutes, the sample would have halved so many times that virtually nothing of the original form would be left for our scientist to measure.

The bottom line is that a radioactive dating method cannot go further back in time than around 10 half-lives. The longer the half-life, the further back in time the dating method can go. Huge efforts are made to keep laboratories ultra-clean and minimize any contamination so as to allow the smallest and oldest samples possible to be measured. With radiocarbon, the dating range is between 40,000 and 60,000 years, depending on the type of material being dated and the detection limits of the laboratory.

When Libby originally measured the half-life of radiocarbon, he calculated it to be just over 5720 years. But radiocarbon suddenly became the new thing to work on and during the 1950s other researchers got in on the act. They came up with a value of 5568 years. This was at odds with Libby's original measurement. This 3% difference in the numbers had quite a big impact on the final age calculated. It was assumed that Libby had made a mistake. The result: the 5568-year half-life was adopted by the scientific community.

Unfortunately, we now know the correct half-life of radiocarbon is 5730 years (Figure 3.2). This is virtually identical to Libby's first estimate. When the mistake was realized, it was thought to be too late to change; too many ages had been calculated using the 5568-year value. As a result, and by a quirk of history, the incorrect value of 5568 years is used. A little unfairly and even more confusingly, it is called the 'Libby half-life'. In practice, as we shall see later, radiocarbon ages have to be converted onto a calendar timescale and the difference in half-life values is corrected for. Fortunately, all labs use the same half-life value. As long as we're only talking about radiocarbon, all ages are directly comparable.

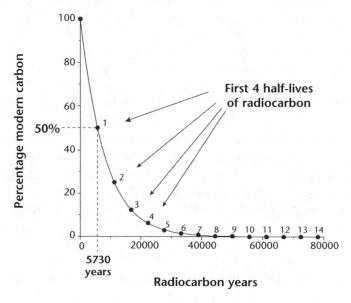

Figure 3.2 The decay curve for radiocarbon
Note: The shape of this curve is identical for all radioactive isotopes

With radiocarbon dating, there are several important assumptions: first, we have to assume that the atmosphere has had the same ^{14}C content in the past as today; second, all things alive have the same concentration of radiocarbon as one another and the atmosphere; and third, that no more radiocarbon is added to the sample after death. In some cases, these assumptions are violated so we have to be careful with what is measured and how the final number is interpreted.

To get a final radiocarbon age, we have to use a point in time to compare against. It's no good saying how old a sample is from when the measurement was made. Radiocarbon dating has been with us for over 50 years. If we were to measure a large ancient seed today that had been analysed early on by Libby, there would be a difference of 50 years, because of the cumulative amount of disintegrations since then. But the plant from which this seed had come could only have lived at one moment in time.

To get over this problem we use AD 1950 as our year zero, and all ages are described relative to this as 'before present', or BP. Say a scientist dated a piece of wood from a tree that grew in AD 950, she'd give it an age centred on 1000 BP. For archaeological samples, BC or AD are often used for convenience.

To complicate matters just a little more, radiocarbon does not give a precise date. Virtually no scientific dating methods give an age to within one year; the exception is tree ring dating, of which more anon. After the measurement of radiocarbon is made in the lab, uncertainties have to be factored into the final age calculation. Rarely is anything perfect: there's always the possibility that a sample has been contaminated in the field or the lab; there can be differences in radioactive decay at the atomic scale; and counting errors with the equipment also have to be allowed for. So an uncertainty has to be given, which gives an age range within which the lab is confident the sample most likely lies.

If we return to our scientist in her lab, we could get her to do an infinite number of measurements on one sample. She'd have to have all the time in the world, as well as vast amounts of money and sample, but let's just pretend anyway. Assuming she hadn't gone mad, our scientist would find that she would have obtained lots of slightly different radiocarbon ages. Not widely different, but enough so that when she plotted them up, they could be added together to form a bell-shaped curve: a normal distribution (Figure 3.3). In a normal distribution, most of the values fall close to the correct age in the middle of the curve, and fall away in number away from the mid-point.

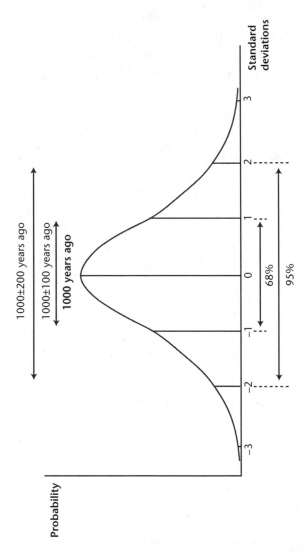

Figure 3.3 The normal distribution

Unfortunately, one single age could fall anywhere on this plot. We have no way of knowing where it would lie unless we actually did undertake this exercise. Thankfully, we don't need to have all the time in the world for measuring the same sample; this variability can be statistically modelled to get an age uncertainty: the standard deviation. In radiocarbon dating, one standard deviation is used as the norm, quoted as '1σ', and gives a 68% confidence level that the age falls within a particular range.

From the wood example above, the sample radiocarbon dated to 1000 BP might have an uncertainty of 100 years. This would be written as 1000 ± 100 BP. We would be saying there is a 68% likelihood that this part of the tree formed sometime between 900 and 1100 years before 1950; or, to put it another way, between AD 850 and 1050. If we wanted to increase our confidence further, we could double the age uncertainty to 1000 ± 200 BP. This would give us a 95% likelihood, or 2σ, that the correct age would lie somewhere between AD 750 and 1150.

⌛

For a long time the Church resisted attempts to radiocarbon date the Turin Shroud, largely because quite a lot of material was needed. The daters would have destroyed much of the cloth. In the late 1970s, a new approach offered hope. Called accelerator mass spectrometry (AMS), and based on nuclear physics accelerators, this method allowed researchers to measure the extremely small differences in the mass of isotopes to count individual radioactive atoms. The results were sensational. No longer were large amounts of material essential. AMS took analysis time down from around 50 hours per sample to a few minutes and needed just a teaspoon of organic material. One gram was often enough for an age. Suddenly, radiocarbon dating the Shroud became a real possibility.

Lots of discussions were had as to sampling and pretreating the cloth. By 1986, seven radiocarbon laboratories submitted recommendations for a protocol to date the Shroud. In 1987, the Archbishop of Turin consulted with the Vatican and selected three AMS radiocarbon laboratories: Arizona, Oxford and Zurich. These were appointed to undertake the work, with the British Museum overseeing the sampling. Sampling took place on the 21 April 1988 in the sacristy of the Cathedral of John the Baptist, with virtually all the process filmed and watched by numerous onlookers. A single 1 cm by 7 cm strip was cut from the Shroud and split into three samples, each weighing about 50 milligrams, an amount that would have been impossible to date pre-AMS. Three similar types of linen were also given to the labs to prepare and measure alongside the Shroud for comparison.

It is worth pointing out at this stage that the radiocarbon age would not represent when the Shroud was used, but when the flax was harvested to be made into the linen. This would have been the time when the last of the radiocarbon was fixed by the plant before harvesting. In itself this wasn't thought to be a problem for dating the Shroud, as it was felt that the cloth would not be much more than a few years old before it was used for burial. Within the likely errors of the technique, a few years difference between harvesting and use would be neither here nor there.

The ages obtained on the Shroud were reported in the journal *Nature* in 1989 and caused a lot of excitement. Arizona reported an age of 646±31 BP, Oxford 750±30 BP and Zurich 676±24 BP. When the errors on these ages were compared, they were found to be statistically indistinguishable from one another at the 95% confidence interval. The values could therefore be averaged together, to give an age of 689±16 BP. The Shroud wasn't 2000 years old.

We mentioned earlier that one of the key assumptions with this dating method is that the amount of radiocarbon in the atmosphere has not changed over time. This is actually not true. The total amount of radiocarbon in the atmosphere does vary, stretching and condensing different amounts of 'radio-carbon time' in the past. The practical upshot of which is that one radiocarbon year does not equal one calendar year. Fortunately, we can correct for this but to do so we need to convert radiocarbon years onto a calendar timescale by using precisely dated wood.

Many species of trees grow by adding a 'ring' onto the outside of their trunk, just below the bark, each year. We'll look at this in more detail later, but suffice to say, by measuring the number of tree rings, one can literally count back through the years and work out the calendar age of the wood. Because trees photosynthesize in the atmosphere, the leaves, and ultimately the rings, record the radiocarbon content of the atmosphere. This is a direct measure of what the ^{14}C concentration was in the air when photosynthesis took place. By repeating this exercise on separate blocks of wood that formed in the past, scientists have worked out how the radio-carbon content of the atmosphere has varied. This has allowed radiocarbon years to be mapped onto a calendar timescale in a plot called the 'radiocarbon calibration curve'. Because of changes in the Sun's activity, the strength of the Earth's magnetic field and our planet's carbon cycle, the radiocarbon content has not been constant. Instead, it's characterized by plateaux, interspersed with times of very rapid change. Sometimes the radiocarbon clock runs slower than 'real' time, during other periods it runs much faster.

Using the latest version of the radiocarbon calibration curve, the Turin Shroud is dated to somewhere between 1275 and 1381. First, this demonstrates that it couldn't be the burial cloth of Jesus Christ; second, it suspiciously overlaps with when the Turin Shroud first appeared in historical records – during the 1350s. It seems de Charney was not as chivalrous as his contemporaries thought. The Shroud was a medieval forgery. Almost before the ink had dried on the *Nature* paper, cries of 'foul play' went up.

For a start, contamination can occur with any radiocarbon sample. Some commentators suggested that over time, the stitching of the Shroud may have been restored or replaced using more recent linen. If the image on the Turin Shroud was really 2000 years old, could it be that the samples of the cloth used for dating were taken from a section that was relatively young? The problem with this idea is that the cloth itself is made up of a distinctive herringbone-style weave. When the protocol for dating the Shroud was first drawn up, the original idea had been to prepare and date samples with the same weave; these would appear indistinguishable to the scientists in the lab. After a world search, no similar woven samples could be found. As a result, anyone with a limited knowledge of the Shroud would know its weave and be able to instantly identify a sliver. Unfortunately, this left the scientists open to the accusation that the sample had received special attention in the laboratory. On the plus side, it did mean that any material that was clearly not linen woven in a herringbone style could be removed, reducing the risk of any contamination.

It was almost immediately commented on that for a short amount of time during the sampling day in 1988, the samples were left with just one individual and not filmed. Could they have been switched? When the samples were investigated under a microscope, the herringbone weave was found to be identical to that of the rest of the Shroud. It would have been extremely difficult, if not impossible, to perfectly reproduce the same weave as the rest of the Shroud.

Alternatively, it has been suggested that bacteria living on the cloth's surface may have contributed enough carbon to skew the age of the linen. Because bacteria fix modern carbon dioxide, when they die they could leave deposits on the surface. The residue could add lots more radioactive carbon to the sample and shift the result to an artificially younger age. Certainly, this is theoretically possible. But for this to happen, 64% of the carbon would need to be modern to shift an age of around 2000 years to the fourteenth century. Such a large amount of bacterial contamination would have been visible to the naked eye. There is some evidence that shifts of up to 400 years can happen when no attempt is made to remove the contamination. Unfortunately for the fanatics, the labs involved had long developed methods to remove contamination and these had been successfully used on literally thousands of other samples. Why would the Shroud be different?

The most creative interpretation of the discrepancy in the ages was based around the observation that the Resurrection was a unique physical event. No one could argue with this. But, Shroud enthusiasts suggested that some of the neutrons in the many atoms that make up a body would have been given off during this event. These neutrons, they reasoned, could then have been captured by ^{13}C atoms in the cloth; turning them into ^{14}C and increasing the radioactivity of the Shroud to give an artificially young age.

Since the intensity of the neutrons produced by such a process would have varied with distance from the body, linen samples closer to the human image should be younger than those reported in 1989. This could be tested by further measurements, assuming permission could be obtained to sample the Shroud again. The problem for this imaginative idea is that so much radiocarbon would have been generated, the ages should have been modern. Instead, all the values obtained were suspiciously close to the first known historical accounts. Ultimately, as the leader of radiocarbon dating team at Oxford University, Robert Hedges, said: 'If we accept a scientific result, we must exercise a critical notion of the probabilities involved. If we demand absolute certainty, we shall have to rely on faith.'

Chapter 4
THE PYRAMIDS AND THE BEAR'S GROIN

Soldiers!
From the top of these Pyramids,
40 centuries are looking at us
NAPOLEON BONAPARTE (1769–1821)

The pyramids of Giza are the only one of the seven wonders of the ancient world to survive. Why and when were they built? Arabic legends dating from the Middle Ages link the pyramids to King Saurid who had a dream that the Earth would turn upside down and the stars would fall to the ground; Saurid took this dream as a prophecy that the end of the world was near and had the pyramids built to house all civilization's knowledge. In Christian Europe, the pyramids were believed to be the biblical Joseph's grain store while he was in Egypt. In these more enlightened days, we now know that the pyramids were tombs of the ancient Egyptian kings and other important officials. Because of this affiliation, the pyramids of Giza are thought to be thousands of years old, but can we get a precise date for when they were built?

To place the historical events of ancient Egypt within our calendar system means we need to translate a whole host of different sources. Probably the best known are hieroglyphics. These began relatively simply, using pictograms to record royal possessions and later came to be used for major commemorative and religious inscriptions. By the time of Alexander the Great's death in 323 BC, the Greeks had introduced the term 'hieroglyphics' to describe this form of writing, from the words *hieros* meaning 'sacred' and *gluphe* meaning 'carving'. By the fifth century AD, Egypt's culture had taken on board so many Christian, Roman and Greek influences that traditional writing had become isolated within temples of the old faith. The last example of hieroglyphs, for instance, is dated to 24 August AD 394 and was written on the small island temple complex of Philae near Aswan in the southern part of the country.

Although less well known, there are actually three other script types of ancient Egypt: 'hieratic' was a simplified version of hieroglyphics, used only in religious contexts; 'coptic' was written in Greek but had several characters from the Egyptian language and used vowels (something that hieroglyphics lacked); and crucially, a shorthand version of writing was also developed called 'demotic', from the Greek word *demotikos* meaning 'popular'. Demotic appears to have survived longer than hieroglyphics. In Philae, a demotic inscription has survived from 2 December AD 452.

Even before the nineteenth century, it was clear that the Egyptian civilization was one of the earliest and greatest. Large numbers of temples and other monuments were found along the Nile, covered in hieroglyphs. The problem was that although these carvings were clearly writing, no one could understand them. Armies of scholars tried to break the code. In 1761, some progress was made when the Frenchman Jean-Jacques Barthélemy correctly realized that symbols enclosed in an oval contained royal names. These ovals were called 'cartouches' because of their similarity to French musket cartridges used at the time. It was later realized that some of the hieroglyphs were alphabetic symbols but no one made any real progress until Napoleon invaded Egypt in 1798.

Napoleon was only in Egypt for a few years. Although thousands of combatants died in the whole sorry saga, there was at least one practical benefit for scholars. In 1799, while extending Fort Julien at el-Rashîd on the western branch of the Nile, a Napoleonic soldier found a black stone covered in

Egyptian text. This was tremendously important – el-Rashîd was anglicized and the find became known as the Rosetta Stone.

Measuring 1.1 m by 0.7 m, the Stone is covered with 14 lines of hieroglyphs, 32 lines of demotic and 54 lines of Greek script. We now know that it records a synod of Egyptian priests in Memphis honouring the young ruler Ptolemy III in 196 BC. The Stone's great importance lay in the realization that it must hold the key to deciphering hieroglyphics. It was considered so important that when the French were defeated in Egypt in 1801, the victorious British demanded the Rosetta Stone as part of the terms of surrender. It can now be seen in the British Museum.

Copies of the Rosetta Stone inscriptions rapidly spread as scholars around the world tried to crack the code. They could smell blood. An Englishman called Thomas Young managed to identify 204 words of demotic and 13 hieroglyphs before giving up with exasperation in 1818. The turning point was in 1822 when the Frenchman François Champollion finally managed the breakthrough. He recognized the name Ptolemy in Greek and demotic on the Rosetta Stone, allowing him to find the cartouched version in the hieroglyphic text. He followed this up by working on the hieroglyphics from Abu Simbel. Here he recognized that the last two identical symbols in one particular cartouche were 'ss'. The first sign in the cartouche was a symbol of the Sun, which he took to be the sun god 'Ra', making 'Ra___ss', the royal name Rameses. A similar name was in a different cartouche, but instead of the Sun there was an image of an ibis, a symbol linked with the god of writing and knowledge Thoth: the cartouche spelling 'Tuthmosis'. Champollion had cracked the hieroglyphs. The story goes that, calling his brother, he threw a handful of papers on the table and shouted 'I've got it!', and then understandably fainted.

⌛

Over the years, armies of archaeologists have travelled across Egypt, many translating the hieroglyphs on any monuments they found. Thanks to the groundbreaking work by Champollion, they have often been able to identify the reigning monarch at the time. As a result, a historical record of Egyptian leaders and other important priests and officials has developed, collectively called the 'king-lists'; often with major events listed against an individual's reign.

The problem for scholars is that, unlike the Romans, the Egyptians didn't date events from one fixed point of time. Instead, the reign of each king was treated as a fresh start, sometimes with good reason. As far as the Egyptians were concerned, the commencement of a leader's reign was a new beginning. Each reign had its own importance. They reasoned that the progression of time did not have to take account of what had gone before. So for a scholar to build up a continuous list of kings, each individual's reign from the different hieroglyphic declarations across Egypt has to be linked to the others. That's a heck of a lot of work. Each individual year of each individual king has to be accounted for, stretching back to the middle of the third millennium BC.

One of the most important records of the different kings is a slab of black basalt called the Palermo Stone. This has hieroglyphics on both sides, recording different rulers from the mythological origins of Egypt up to around 2400 BC. Another key element is a 'history' of Egypt that was written by a priest called Manetho in the third century BC, allegedly stretching back to around 3100 BC. Unfortunately, the original text no longer exists: we only have fragments of Manetho's history copied by later writers and travellers. The rest of the king-lists are made up of fragments preserved on tomb walls and other reliefs covered in hieroglyphs.

When the Egyptians were recording the years of their kings' reigns, they were working to a 365-day calendar, probably calculated from the annual flooding of the Nile on which their civilization depended. A year comprised 12 months of three weeks or decands, with each decand having 10 days. This makes a total of only 360 days, so another five were tagged on at the end of the harvest season to make a full year. Although the Egyptian calendar was described by the great Austrian mathematician Otto Neugebauer as 'the only intelligent calendar which ever existed in human history', over the long term, the absence of six hours in each year was enough to introduce a significant offset between the calendar and the seasons. We know what problems the Romans had.

A key part of linking the king-lists to today is through astronomical observations that can be independently dated. Crucial for this is the star Sirius, also called the Dog Star: this was known to the Egyptians as Sopdet and was one of the brightest lights in the night sky. Originally, for the Egyptians, its rise on the horizon just before sunrise coincided with the start of the Nile flood and marked the beginning of the calendar year. Even as early as 3000 BC, the goddess Sopdet is shown in an inscription as a seated cow with a plant between her horns; a symbol used in hieroglyphics to mean 'year'.

A problem for the Egyptians was that the cumulative shortfall of six hours a year meant the rising of Sopdet only coincided with the start of the 365-day Egyptian administrative calendar once every 1460 years: the Sothic cycle. Thankfully, when the two coincided in AD 139, the Romans were in control of Egypt and celebrated with the issue of a commemorative coin. As a result, we can back-calculate the date of the other observations when the rise of Sopdet coincided with the start of the Egyptian calendar; sometime around 1321–1317 BC and 2781–2777 BC. These astronomical events were recorded during specific reigns and give some critical reference points for linking the king-lists to our calendar.

Unfortunately, linking Sopdet to the calendar might not be

as simple as it first seems. Historians have tended to assume that astronomical observations were made at Memphis or Thebes in the middle part of the River Nile. Yet depending on which latitude the measurements were made, different dates can be calculated for the conjunction of the rise of Sopdet and the start of the Egyptian calendar. It is possible that the observations were made at Elephantine to the south, or elsewhere. Festivals celebrating this coincidence would therefore have been made at different times, depending on where the observations were made in the country.

What seems amazing is that despite all their achievements, the Egyptians decided to ignore the long-term divergence of their calendar from the seasons and continue with a fixed year of 365 days. The Egyptians could see it did not track the seasons. After all, they went through several Sothic cycles over the millennia. Perhaps it had a cultural significance lost on us today. Whatever the reasons, they carried on using a calendar that bore no relationship to the passing of the seasons for thousands of years. In 238 BC, during the Greek Ptolemaic period, the leap year was introduced but largely ignored until Augustus put his foot down and insisted it was used in 30 BC.

So what do we have so far? The Egyptians had a civilization that spanned millennia but they treated each individual king's reign as a fresh start and so kept no continuous records of all the kings who had reigned. Instead, we have a handful of ancient compilations, with records of individual reigns preserved across the country in the form of hieroglyphs. Just to add to the confusion, the Egyptians used a calendar that had no leap year. To get a historical date for the construction of the pyramids, the list of kings who reigned has to be somehow tied into the modern calendar.

There were 31 dynasties in Egypt, each comprising several kings. They lasted until the suicide of Cleopatra VII and the murder of Caesarion, her son by Julius Caesar, in 30 BC, which resulted in the integration of Egypt into the Roman Empire. Most of the dynasties formed 'kingdoms' – stable periods in which it is relatively easy to plot the line of kings with their dates of reign.

The big problem comes with the 'intermediate periods': unstable times when Egypt experienced major events such as invasion, civil war and famine. In the worst cases, all these disasters happened at the same time and often resulted in the collapse of the state into several petty kingdoms, each led by their own king. These are a major headache for scholars trying to identify the relationship between the various monarchs and the period of their reign. In some cases, independently dated astronomical observations have given a chronological fix. Unfortunately, there aren't enough of these to go around to get a precise series of dates for all the kings.

Because of these uncertainties, there are now several different versions of the king-lists. These vary in the timing and length of reign of individual leaders, which together can add up to several centuries of difference. This is quite a problem if we want to try to understand who was building what in Egypt and how this relates to other events in the region. To some extent, you can choose any date you like.

Radiocarbon dating of archaeological finds could be attempted to get us around this apparent impasse. Yet, as we saw with the Turin Shroud, depending on past changes in the atmospheric radiocarbon content, the errors of tens to hundreds of years would probably be the same as – or in some situations worse than – the historical uncertainties. Even if a precise radiocarbon age could be determined, it would only give an indication of when a site was used and not when it was erected. The construction date is crucial if we want to unambiguously link a site to a historical figure.

⧖

While the sheer size of the pyramids is a phenomenal achieve-ment, equally remarkable is just how precise their orientations are. The Fourth Dynasty Great Pyramid of Khufu, also known as Cheops, is 230 m on each side, 147 m high and made up of around 2.3 million blocks of stone, each weighing about 2500 kg. The sides of this and many other pyramids point almost precisely true north. In fact, the sides of the Great Pyramid are only three arc minutes off – one arc minute being just $\frac{1}{60}$ of a degree.

How could an Egyptian architect have managed such a high level of precision when laying out the ground plan for a pyramid several thousand years ago? Assuming he had a clear view of the horizon, our architect might have tried taking the halfway point between where the Sun rises and sets. The problem is that taking measurements of objects on the Earth's horizon is notoriously difficult – largely because of the inter-ference from the atmosphere; it's just not possible to get within three arc minutes of true north using this method.

Intriguingly, before and after the reign of Khufu, the pyra-mids were not so accurately aligned. It seems odd that once the method of finding geographic north had been cracked in Khufu's reign, the pyramids were not thereafter equally precisely aligned.

In 2000, Egyptologist Kate Spence at Cambridge University put forward a fascinating explanation as to why the Great Pyramid was almost precisely aligned to true north and the others were not. To follow her explanation, we need to look at how the Earth orbits the Sun.

Over a human lifetime, the way the Earth rotates around the Sun can be thought of as pretty much unchanging. The Earth rotates at an angle of 23.5° from the vertical and travels around the Sun in an elliptical orbit. The extreme positions of summer and winter mark the times in the Earth's orbit where one of the hemispheres is directed towards or away from the Sun. Between these points, the equinoxes mark the times of the year when both hemispheres are at right angles to the Sun, with the result that day and night are of equal length.

We have seen that in AD 325 the Council of Nicea agreed to calculate Easter relative to the vernal equinox – arbitrarily set at March 21. Orbitally speaking, this is not strictly correct. The vernal and autumnal equinoxes are not actually fixed in time. From a northern hemisphere perspective, these equinoxes drift a few days either side of March 21 and September 23 because the year is not made up of an even number of days. Anyone can notice these changes in a lifetime.

Over thousands of years more significant changes take place in our planet's orbit around the Sun. Because of the gravitational pull on the equator by our star, the Moon, and the other planets, the Earth's rotation experiences what is commonly called a 'wobble'. This can be best illustrated by imagining the axis of the Earth's rotation extending out from the North and South Poles far into space. Over time, the axis traces out a cone, like a gyroscope or spinning top. The upshot of all this wobbling is that it changes the Earth's orientation as we orbit the Sun. The orbital position of the equinoxes and all the seasons relative to the Sun shifts, and this is known as the 'precession of the equinoxes' (Figure 4.1).

The important point to take from all this is that the axis through the geographic North and South Poles points towards different parts of space over time. After 26,000 years, it returns to the same area in the night sky. This innocent-sounding change plays quite a significant role in allowing us to date the pyramids.

The precession of the equinoxes has a major impact on the celestial pole. This is the part of the night sky around which the stars seem to rotate. At the moment, Polaris, also known as the Pole Star, is at the northern celestial pole. Regardless of what time of night you go out, Polaris always appears to be rigidly fixed over geographic north, and all the constellations move around it. But Polaris has not always been at the celestial pole. In many ways, we're extremely fortunate it's where it is now, given how handy it is for navigation. As early as 130 BC, the Greek astronomer Hipparchus of Nicea noticed that the celestial pole changed over time when he compared his observations with those made earlier by the Babylonians.

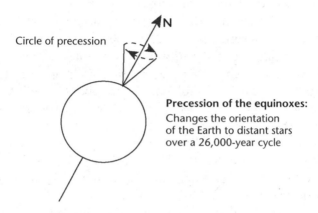

Figure 4.1 The 'wobble' in the Earth's rotation causes the precession of the equinoxes

An excellent example of how the precession of the equinoxes might impact on our lives is through the zodiac. The Babylonians had been among the first to join up certain stars to trace out constellations; this supplemented their calendar and was, they believed, of astrological significance. By around 500 BC, the zodiac had reached the form we now recognize. They divided the night sky into 12 segments, each defined by the constellation that rose in the east immediately before the Sun. Because of the precession of the equinoxes, Hipparchus recognized a slow westward shift in the constellations. The result: at the time of Hipparchus, the constellation

Aries rose with the vernal equinox; for the last 2000 years Pisces has had this honour; and relatively soon Aquarius will assume the role. The dates of the zodiac signs used for astrology, however, were set at the time of the Romans. Because of the precession of the equinoxes, their dates are hopelessly out of track with today's calendar. If you want to believe in astrology, you need to check the predictions one star sign earlier on the zodiac charts.

Anyway, let's return to our Egyptian architect. He could have used the celestial pole to align the sides of the pyramids. He might have built a platform for a plumb line, and then using a heavy weight on a piece of string, lined up against the celestial pole. The only problem was that because of the yet-to-be-discovered precession of the equinoxes, Polaris would not have been there. What was there at this time? Inexpensive computer software can show the night sky at any location in the past and future. Running this for the time of the kings in the Fourth Dynasty, we find literally nothing. No star was at the celestial pole.

Spence suggested that just because there was no Pole Star at this time, our Egyptian might still have used the concept: two bright stars falling on a straight line either side of the celestial pole would have done the same trick. Our computer software for the night sky gives us two possible combinations of stars for the Fourth Dynasty. The brightest and most likely of these is Kochab (in the constellation known as the Little Bear or Little Dipper) and Mizar – from the Arabic word meaning 'the groin' – (in the Great Bear of which the Big Dipper forms a part). One other possible star combination is not so obvious to the naked eye; we'll come back to this later.

Our ancient astronomer could have used a plumb line when both stars were vertical to the Earth's surface. This would have given an accurate fix on the location of geographical north. If this is done with Mizar and Kochab during the Fourth Dynasty, a date of 2467 BC can be calculated. But we know that the

Great Pyramid sits just west of geographical north. Over time and with the precession of the equinoxes, Mizar and Kochab would have tracked over the celestial pole from the west. So, assuming our Egyptian had a steady hand when he was measuring for the alignment, an offset of three arc minutes to the west of north would give an age of 2478 BC (Figure 4.2).

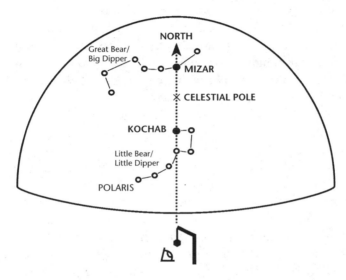

Figure 4.2 Making the alignment for the Great Pyramid of Khufu against Mizar and Kochab in 2478 BC

So when in a king's reign did the alignment for the pyramid take place? There wasn't much point in completing the measurements at the end of their stint on the throne. It's estimated that around 30,000 people would have been needed to build the Great Pyramid; it's unlikely a successor would have wanted to spend so much time and resources glorifying his predecessor's reign. Far more likely is that the construction would have begun at the beginning of the reign, possibly in the second year. Using this approach, Khufu can be dated as coming to the throne in 2479 BC. The historical king-lists vary concerning Khufu's ascension to the throne. He was the second king of the Fourth Dynasty. A consensus 'date' of around 2554 BC would suggest the lists are around 75 years too old.

Although the alignment against the celestial pole during Khufu's reign is a fascinating suggestion, it might be a lucky hit or just plain wrong. After all it is just one pyramid. The convincing part of this dating method is when we try it out on the other pyramids. Remember, those built before Khufu were aligned too far west of geographical north, while most of the later constructions were too far east.

Snofru, who was also known as Sneferu and Snefru, is an excellent example. He had the first pyramid built at Meidum and led Egypt immediately before Khufu. Sadly, most of the pyramid collapsed some time after construction but even today it still has a tremendous sense of grandeur. Snofru's Meidum pyramid has a western side 18 arc minutes west of geographic north. Using the traditional historical dates, the accession of Snofru is supposed to have been 2600 BC. If we use the same trick as the Great Pyramid, we get a new ascension date of 2526 BC. This is a difference of 74 years; virtually identical to Khufu. The method looks hopeful.

If we go to the other side of Khufu in time, we can focus our efforts on the Fifth Dynasty pyramids built at Abusir, south of Giza. Unlike the Fourth Dynasty, however, the pyramids built by the next lot tend to be ruins. Somehow they seem to have lost some of the skills in pyramid building that their predecessors had had. In contrast to most of his royal line, Neferirkare had a pyramid built in a similar stepped shape to that of the Fourth Dynasty. Perhaps he hankered after the good old days. Assuming his astronomer still knew the tricks of the trade, the alignment 30 arc minutes too far to the east gives an ascension date of 2372 BC. The traditional age was 2433 BC, a difference of 61 years. Pretty close to a constant age offset between the two methods of dating.

Strangely, the odd pyramid seems to be aligned completely differently. The pyramid of the second king of the Fifth Dynasty, Sahure, was built sometime after Khufu's. Traditionally, he was thought to have ascended the throne around 2446 BC. If this is the case and Spence is right, why would his pyramid be 23 arc minutes too far to the west? Surely the alignment should be too far to the east? This seems to be a major blow to Spence's theory. Or is it?

Let's just hold one thought: with the changing wobble in the Earth's spin, a plumb line drawn between Mizar over Kochab will be slightly west of geographic north before Khufu's reign; afterwards, the two stars will be to the east. Although it is true that Mizar can be over Kochab in the night sky, this is only the case for half of the year. In the other half, the opposite happens: Kochab lies over Mizar. The same shift in arc minutes from north would still happen but it would be in the other geographical direction. If some of our ancient Egyptian astronomers made the alignment for the pyramids six months different to most of their colleagues, the offset would appear to be in the opposite direction to the main trend.

We can correct for this and plot all the pyramids using the difference from geographical north in arc minutes, regardless of whether it's west or east. When we do this all the pyramids fall on a straight line (Figure 4.3). Surely this is too incredible a coincidence to be just chance?

We mentioned earlier that one other pair of stars might also have allowed the Egyptians to locate the celestial pole at the time of Khufu. These are ε-Ursae Majoris and γ-Ursae Minoris, two relatively faint stars in the same constellations as Mizar and Kochab. This alliance of stars would 'predict' a date for the onset of Khufu's reign at 2443 BC. This is not bad. It's fairly close to the historical date for the start of Khufu's reign

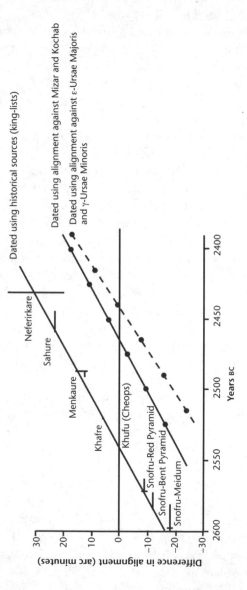

Figure 4.3 Dating the Egyptian pyramids of the Fourth and Fifth Dynasties

in 2554 BC. But if our ancient astronomers had used this particular combination of stars on the other pyramids, the method doesn't work as well: the ages become even younger than those calculated using Mizar and Kochab. This would mean there's a far bigger problem with the historical records than we originally thought. Not only that, they do not track the changing alignment of the different pyramids with a constant offset. The trend converges on the historical ages into the Fifth Dynasty. The gradient becomes very steep compared to the historical trend (Figure 4.3). This would mean that many of the errors in the historical dates lie in the Fourth and Fifth Dynasties: pretty unlikely considering this was a stable period when one king assumed control after another.

It therefore seems probable that the ancient Egyptians used the stars in the Little Bear and Great Bear's groin to align their pyramids. By working through the method, these amazing constructions can be dated to within five years as far back as 4500 years ago. This is almost more accurate than many of us can remember events in our own lives.

Chapter 5

THE VOLCANO THAT SHOOK EUROPE

Time and tide wait for no man
ENGLISH PROVERB (FOURTEENTH CENTURY)

Santorini is one of the most romantic islands in the world. Nestled down in the eastern Mediterranean, it's a regular stopping-off point for cruise ships exploring the beauty of the Greek islands. Strictly speaking, it's not just one island but a collection, together making up a doughnut-ring shape. Sections of the northern and southern ring are now gone, allowing the sea water to flood into the central basin which is an impressive 84 sq km. The largest island, Thera, makes up the eastern, northern and southern sides, the view of which from the centre is one of most spectacular sights of the natural world: a sheer cliff of different coloured rocks reaching a height of up to 300 m above the sea; the town of Fira seemingly splattered onto the cliff face. Sadly, the only time I've visited it was for work and not with my wife. Some things are never forgiven.

Santorini has a long history of volcanic eruptions. Over the last 1.6 million years, this island has regularly spewed out vast amounts of rock, covering itself with yet more layers of different coloured debris that together make up a continuous record of its eruptive history. Although many of these eruptions have been enormous, a major reason why Santorini has been almost ritualistically prodded and explored by scientists is the impact that the last major eruption may have had on the neighbouring island of Crete around 3500 years ago. This eruption was truly enormous, with the volcanic column probably reaching over 35 km in height. There is still a lot of argument about its size but a conservative estimate puts it that the volcanic material produced would have been enough to cover the whole of Western Europe with a 1-cm-thick layer.

A relentless stream of television documentaries wax lyrical about the collapse of the first European civilization, the ancient Minoans, based on Crete several thousand years ago. The TV shows almost always say the same: the Minoans were remarkably advanced for their time; they traded on an equal footing with the other local 'superpower', the Egyptians; they had colonies throughout the eastern Mediterranean; and then, mysteriously, they disappeared, seemingly overnight. Even now it sounds like a ripping good yarn. But invariably, the programme makers tend to make out that they have discovered groundbreaking evidence that the demise of the Minoans may be even more spectacular than previously thought: an eruption of Santorini, 120 km north of Crete, could have been the cause of the civilization's collapse. If the director is really getting carried away, they'll often get a shot of an academic sitting on the beach, looking pensively out to sea. No wonder the poor saps often look uncomfortable: the idea has been kicking around for over 60 years.

In a 1939 issue of the archaeological journal *Antiquity*, the Greek archaeologist Spyridon Marinatos suggested that the eruption of Santorini may have caused the collapse of the Minoans. Marinatos was one of the greats in Greek archaeology and in many respects ahead of his time, drawing on observations from the 1883 eruption of the Indonesian volcano Krakatoa. This was about a third of the size of Santorini, yet was heard 4600 km away and created a series of giant waves. Marinatos suggested that it was not the direct blast of the eruption but the associated effects that may have laid waste to Crete. Thick layers of ocean sediments, ash and pumice are found splattered across coastal areas throughout the eastern Mediterranean, including the north coast of eastern Crete.

Marinatos proposed that a tsunami swept south from Santorini, devastating coastal populations on northern Crete and irrevocably destroying the Minoans' maritime power base, weakening their civilization. Under this scenario, Crete would not have been a good place to be at the time. Marinatos reckoned – with surprising accuracy – that the date of the eruption was around 1500 BC. The editors of *Antiquity* took the unusual step of putting a short note at the end of the paper to the effect that, although interesting, the ideas needed testing before they should be accepted as fact. This caveat set the tone for the debate through to the present day.

⋈

Before the twentieth century, stories of King Minos and a Cretan society were largely thought of as myth. The great historians Herodotus, Homer and Thucydides describe a strong maritime civilization called the Minoans who were based at Knossos on the island of Crete. They were well-enough organized to have the first naval fleet in the region, allowing them to drive out pirates and link together their many colonies through the eastern Mediterranean.

It was only at the beginning of the twentieth century that it looked as if such stories might be true. In 1878, a local Cretan, appropriately named Minos Kalokairinos, started digging in a large mound just outside Heraklion, near the central north coast of Crete. He discovered what was later found to be part of a throne room and some palace store-rooms. Unfortunately for him, the Ottoman authorities at the time would not grant him a licence to excavate the site. In the late 1880s, the famous German archaeologist Heinrich Schliemann, who claimed to have discovered Homer's Troy, believed the site was the home of the legendary King Minos. The story goes that he was so annoyed at the Turkish landowner for exaggerating the number of olive trees on the site that he refused to purchase the land, and so didn't complete any excavations. Only after Crete gained independence from the Ottoman Empire did the British archaeologist Sir Arthur Evans get permission to excavate the site in 1900, after he had made contact with Kalokairinos.

What Evans discovered was beyond anything imagined. The mound was found to contain a sophisticated, extensive Palace, with the technology capable of delivering clean water to at least 2000 people, and surrounded by a town with a population several times that size. Based around the worship of a bull deity, the Minoan horns are seen thoughout Crete. The site even had the first European theatre and paved road. The international media frenzy made Evans a household name virtually overnight. During the course of the excavations, Evans made 'restorations', many of which sparked controversy; but they do give an excellent impression of what the Palace may have looked like.

The Minoans were a remarkably advanced civilization. At the centre of a maritime trading empire as early as 2000 BC, they travelled throughout the eastern Mediterranean. We now know there were centres of Minoan population almost everywhere in the region: the Greek islands and mainland; the Levant; and also Egypt. Soon after Evans' finds on Crete, it was realized that the Minoan culture produced vast amounts of distinctive-styled pottery including bridge-spouted jugs, stirrup jars and stemmed cups. They appeared to have got everywhere. What the Cretans made seemed to be something neighbouring cultures just had to have. Not long after, it was realized that the style of the pottery was not all the same. It seemed to change over time. Fortunately for archaeologists, this evolution of artefact style can date finds: a technique called 'typology'.

Ⴠ

Many people are old enough to remember using coins with different heads of state. If the leader hangs around long enough, different busts are used as the individual gets older. In the UK, many remember decimalization; in Europe, the introduction of the euro. The different coin styles before and after these events are instantly recognizable. Almost without looking at a coin closely it is possible to tell how old it's likely to be. If uncertain, you can check with the date struck on the coin. These changing styles over thousands of years were recognized as early as the medieval period. Coin collections were printed from the sixteenth century. But it was not until the end of the nineteenth century that it was thought that these ideas could be applied to other types of artefact.

The first person who recognized the potential of typology was Augustus Lane-Fox, also known even more splendidly as Pitt Rivers. Lane-Fox used the principles of gradual change laid out in Charles Darwin's *Origin of Species* to recognize evolutionary styles in his collections of artefacts. Travelling the Empire as a grenadier guard in the British army, he acquired an enormous collection that ranged from boomerangs to spears and shields. Lane-Fox argued that the greater the complexity, the more an artefact type had benefited from development and modification: it must be relatively young. Simple designs had to be old.

The first successful attempt at using typology to date anything was where there were clear changes in style of one type of material. An unassuming Swedish scholar called Oscar Montelius became the 'father of typology' and focused on the Bronze Age, arranging the artefacts from this period on the basis of the degrees of similarity and dissimilarity. It made him a household name in his home country. He even got onto a stamp.

The Bronze Age falls between the Neolithic, also known as the New Stone Age, and the Iron Age. Inspiringly named, this period was when bronze was the main material type for making tools. When it took place varies depending on when the technology was developed or imported into a region; in

Europe and the Middle East, it seems to have started around 4000 years ago. Using tools and weapons that he found in museums and private collections, Montelius divided this period up into six phases, characterized by distinctive styles and shapes. In 1885 he published a book of his ideas called *Dating in the Bronze Age with Special Reference to Scandinavia*.

A big test of Montelius's ideas was to excavate archaeological deposits. The simplest artefacts should be the deepest and oldest. Many archaeologists felt this approach was far too simplistic and set out to challenge it. But Montelius's ideas held up and soon the principles were being used across Europe. Although modified since then, his chronology is still used to interpret archaeological finds in the region.

☒

In the eastern Mediterranean, pioneering archaeologists working in the Aegean soon realized that the wealth of different pottery remains across the region offered the chance to date the Minoans. This really started at the end of the nineteenth century with the great eccentric British archaeologist Sir Flinders Petrie. In the 1890s, Petrie found Minoan pottery in an Egyptian site called Kahun. Archaeologists became excited. By finding Minoan pottery in Egyptian contexts they could be tied into the king-lists and given historical dates, at least in theory.

During the course of their civilization, the Minoans had four distinct cultural periods. They were not blessed with good fortune, and it appears that at the end of each of these periods, they suffered a natural disaster of such magnitude that their society had to recover almost from scratch. Based on the traditional links to ancient Egypt, these periods have been tentatively dated as follows:

- Pre-Palace Period (2600–1900 BC)
- First Palace Period (1900–1650 BC)

- Second Palace Period (1650–1450 BC)
- Post-Palace Period (1450–1100 BC)

The Minoans reached their peak in the Second Palace Period, and it was at the end of this that Knossos was largely abandoned. When Marinatos suggested a cause for Minoan collapse, he was actually referring to the end of the Second Palace Period. But are the dates of 1500 BC traditionally associated with the Santorini eruption and 1450 BC for the end of the Second Palace Period accurate? Or is there enough uncertainty in both to make it possible that the two happened at the same time? What's clear is that the scientific and archaeological community are almost as divided now as they first were when Marinatos made his suggestion in 1939.

Comfortingly, we know that if we look at the Egyptian king-lists, there isn't a major error in the dating scheme. Importantly, the dates for the eruption and the collapse of the Minoans are both based on links to Egypt. Marinatos, for instance, knew of Minoan sites where volcanic ash and pumice from Santorini had been found, as well as ones with Egyptian pottery. But an error in the king-lists would affect both the date of the eruption and the Minoans' cultural collapse to the same extent. Any problem with dating the Egyptian king-lists wouldn't bring the different dates closer.

One possibility that might explain the difference in ages is that the links to the Egyptian chronology were wrong. Pivotal to all this were the changes in Minoan pottery style at the time of the eruption and the end of the Second Palace Period.

During excavations in Crete, Marinatos found two styles of Second Palace Period artwork. One had pots, vases, jugs and cups covered with horizontally banded decorations and spirals or floral designs. Another seemed to be inspired by the ocean, with many images, such as of octopuses, covering the whole vessel. Originally it was thought that the Minoans were producing these two different styles alongside each other but later excavations found that these were in fashion at disparate times. It seemed that the Second Palace Period could be split into two phases: an 'early', horizontal-banded group, and a 'late', ocean-inspired group.

On the southern part of the main Santorini island of Thera, a major Minoan settlement called Akrotiri has been slowly excavated from the volcanic deposits since 1870. In 1967 Marinatos started work there in the hope of finding support for his idea of a volcanic eruption for the end of the Second Palace Period. Marinatos died in 1974 but the work has continued. The site now measures around 150 m across and seems to represent just a small part of what must have been a substantial settlement.

The preservation of Akrotiri is remarkable, considering it was only 8 km from where the centre of the eruption is believed to have been. It was not destroyed but completely buried in volcanic ash, pumice and boulders over 2000 years before the eruption of Vesuvius in AD 79 that devastated Pompeii and Herculaneum. Impressive frescoes have been found at the site. Lots of the houses contain jars, benches and stone mills, similar to those still used on the island today. Many of the houses are two- and three-storeys high, showing the Minoans were skilled builders.

Unlike Pompeii or Herculaneum, no bodies, valuables or food have been discovered. There was almost certainly enough earthquake activity before the eruption to give people time to leave. In one building, there was even enough time to remove a set of three beds from the ruins of an early earthquake and pile them up. Where the inhabitants got to is unclear, but it is unlikely they managed to get away in time. Maybe one day in the future, excavations will find the population buried on the shore as they desperately waited for ships to take them to safety.

Critically, lots of early phase designed pottery has been

found at Akrotiri, including two vases that have almost assumed godlike status. These vases were decorated with the forerunner of another late phase style, a double axe-head design. But no fully developed late pottery has ever been found in Akrotiri. In contrast, on the Minoan stronghold of Crete, lots of both design types have been found.

By 1980, Santorini ash was discovered on the Greek island of Rhodes, in sequences that were clearly deposited before the late phase. This was soon followed by finds on Crete itself, where ash was found in excavations that were clearly early. This pegged down the relative position of the eruption to the very end of the early phase of the Second Palace Period. But what was the date of the eruption?

Marinatos had originally suggested 1500 BC as the date for the eruption. This was based on just a handful of Minoan and Egyptian finds that could be linked to the king-lists. The problem is that typology is not a precise science. To illustrate this, just consider your parents for a moment. You may have been lucky and your relatives were trendsetters, wearing fashions before they were popular. Or perhaps your parents caught on late and persisted in wearing styles long after everyone else had switched to something new. Either way, the important point is that fashionable items don't span a convenient block of time. So when dealing with just a handful of archaeological finds, it's easy to see how you might get the leftovers of a trendsetter or a die-hard hanger-on. The result: your chronology can be wildly off. By the late 1980s, more Minoan and Egyptian finds came to light: the new links showed the dates needed to be shifted further back in time. But by how much?

Early on, archaeologists had tried radiocarbon dating old Minoan sites that pre-dated the Second Palace Period. It was felt that there wasn't much point with the younger sites as these could be dated precisely using the Egyptian scheme via typology. It was found that the radiocarbon ages were older than the supposed historical dates in the pre-Second Palace

Period. This was soon explained away: it was believed the Egyptian scheme was on shaky ground this far back, so a difference in dates was to be expected. We now know this is rubbish.

One way to bypass all the typological uncertainty was to date the eruption directly. From the 1970s onwards, radiocarbon dating was finally being tried on Minoan sites with Santorini material in them. But the age difference between the radiocarbon and historical dates persisted. Instead of dates of around 1500 BC predicted by the links to the Egyptian chronology, 1600 BC and older were being suggested by radiocarbon. Could this be true? To confuse matters more, other techniques were being developed that seemed to give a third result.

Using changes in the Earth's magnetic field, scientists in 1984 analysed the magnetic orientation of grains preserved in pottery. They suggested that the end of the Second Palace Period took place at different times across Crete. A whole range of different camps now formed around what all these results might mean: the radiocarbon dates were systematically contaminated; the calibration curve was wrong; or some other mistake had been made. In some instances, apparently conflicting results were simply ignored. Hardly anyone now refers to the magnetic work. Some data can be too difficult to explain away.

The debate pottered on through the 1970s and early 1980s until work led by American researcher Valmore LaMarche added a whole new dimension to the apparent chaos. Ironically, the new data did not come from the Mediterranean. Instead it was based on tree ring analyses done in the White Mountains of North America using the longest living trees in the world: the bristlecone pines.

In many parts of the world, tree rings can be counted back in time to give a record of individual years spanning thousands of years. It's the most precise and accurate dating method available. As with all dating methods, there are potential problems that we'll discuss later, but, importantly, a tree ring specialist can date an event to a single year. The basic premise of the method is that distinctive patterns of tree ring thickness can be interpreted in terms of the changing growing conditions during a tree's lifetime. If the environment is warm and moist, a tree can grow relatively quickly and produce a thick ring. If the growing season is poor, such as when it is too dry and cold, the tree will struggle, resulting in a narrow ring. If the conditions are extreme enough, no ring will form at all.

In 1984, LaMarche's team reported that they had discovered an unusual period of narrow tree rings starting from 1628 BC. The group argued that the large size of the Santorini eruption must have had global consequences: the release of ash particles and sulphate gas during the eruption would have reflected the incoming Sun's rays and resulted in the northern hemisphere cooling. This finding was soon supported in 1988 by puny Irish tree rings. Researchers at Queen's University Belfast, headed up by Mike Baillie, found a similar collapse in Irish bog oak growth at the same time: 1628 BC.

Meanwhile, researchers working on the Greenland ice appeared to back up an earlier date for the eruption. Counting down through the annual layers of ice, they found a huge amount of sulphate had been laid down around this time. Could this be a volcanic eruption? Perhaps the smoking gun pointing to Santorini? The problem was that the age was even older than the trees. Originally dated to 1390 BC, the new ice core data placed the eruption at 1645 BC. Sceptics moved in quickly. They immediately questioned the conclusions: a global cooling could have been caused by anything, not just a volcanic eruption; the sulphate peak could have been caused by any volcanic eruption, not just Santorini.

The most recently dated twigs and seeds burnt in Akrotiri during the Santorini explosion give a radiocarbon age of around 3355 years ago. Calibrating this to calendar years gives a mean date of 1650 BC. This is a lot older than the original archaeological date of 1500 BC but still has several decades of uncertainty. The bottom line is that no matter how many single radiocarbon ages are made, the uncertainties will always be large because of the shape of the calibration curve.

In theory, the changing shape of the radiocarbon calibration curve can actually be turned to an advantage (Figure 5.1). It is known precisely how radiocarbon has changed in the atmosphere every 10 years over the last 12,000. If a tree could be found that was killed by the eruption, it would be possible to sample consecutive 10-year blocks of wood from the outside of the trunk to the centre. These could then be radiocarbon dated in the laboratory. Because Libby showed that the radiocarbon content of air is identical around the world (Chapter 3), the pattern of radiocarbon ages made from our burnt tree could be matched to the shape of the wiggles in the calibration curve. Much like a complicated jigsaw puzzle, the shape would only fit onto the calibration curve one way. To get a precise date for the death of the tree (and therefore the eruption), we could then focus on where the outermost ^{14}C age measurement falls on the curve. This final age would bypass all the uncertainty and give a precise date because the samples closer to the centre of the tree wouldn't let it move anywhere else in time. Years of expeditions have tried and failed to find such a trunk of wood.

This approach has been tried in Anatolia in Turkey, although none of the wood was burnt in the eruption. Lots of burial chambers have been found, many built by the Phrygians on the central Anatolian plateau. The burials are truly spectacular and the chambers are made of large tree trunks. These trunks have been systematically radiocarbon dated using 10-year blocks to get precise ages for the construction of the burials in the region. The main part of the tree ring chronology for Anatolia comes from just one burial chamber: the Midas Mound Tumulus at Gordion. This is the oldest standing wooden building in the world, at over 2500 years, and in it a large growth spurt has been found, recorded by the tree rings. The Anatolian trees would have been immediately downwind of the Santorini eruption and the ash could have provided much-needed nutrients, allowing them to grow far above their average. The date of the growth spurt is 1645 BC. Could this be the Santorini eruption?

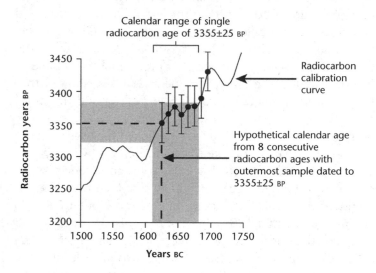

Figure 5.1 Using radiocarbon wiggles to date the Santorini eruption

Commentators regularly emphasize that tree responses to volcanic eruptions are uncertain. Although trees respond to changing growing conditions, what causes them is often not known. All that can be said is that the Midas Mound Tumulus data are consistent with the effects of an eruption. The ultimate test would be to find particles of Santorini volcanic ash in an ice core layer of the right vintage. This would wrap everything up nicely.

Using volcanic ash to link different sites is called 'tephrochronology'. It has been used for dating since the beginning of the twentieth century. The method relies on being able to pinpoint unique characteristics of individual volcanic eruptions. Bands of volcanic ash invisible to the naked eye can now be found in ocean and land sediment cores, providing an incredibly powerful dating tool, often over large regions. A key feature used for tephrochronology is the geochemical makeup of individual ash shards; this is a snapshot of the average composition of the magma during the course of an eruption.

In 2003, after years of frenzied searching, shards of volcanic ash were finally reported to have been found in the 1645 BC layer of Greenland ice. In a book on the proceedings of a conference, researchers claimed that they had the final piece of the jigsaw and the true age of the eruption was known at last. The date for Santorini was fixed.

As is so often the case with Santorini, things were not as clear as they first seemed. The geochemistry that was reported from the ice could not have been more different to Santorini. It was nothing like it. How the correlation was made and published seemed extraordinary. Follow-up work soon showed that the shards were more likely to have come from an Alaskan volcano called Aniachak that seems to have erupted around the same time.

All this leaves the date of the Santorini eruption still up for grabs. We now know that it did *not* coincide with the end of the Second Palace Period, although it is still possible that the eruption did severely damage the Minoan civilization in the long term. The eruption was clearly not 1500 BC and appears unlikely to have been in 1645 BC, unless Santorini did erupt in the same year as Aniachak; which is not impossible. It is odd that there was a tree ring growth spurt in Anatolia, as this could not have been caused by an eruption in Alaska. The alternative date of 1628 BC may be correct but, as we will soon see, something else may have caused the cooling of the atmosphere evidenced by the narrow rings in the American and Irish trees. This all begs one crucial question: will there be another television documentary asking whatever happened to the Minoans?

Chapter 6

THE MANDATE FROM HEAVEN

Thus the whirligig of time brings in his revenges
WILLIAM SHAKESPEARE (1564–1616)

'Can't see the wood for the trees' couldn't be a more real danger for tree ring dating. Where most methods come up with a broad age range for an event in the past, the trees give a date to a single year. This level of precision can sometimes be overwhelming. And it's all down to the simple principle that each year most trees grow by one ring. Staggeringly, this realization goes back to the dawn of scientific thinking: Theophrastus, Greek philosopher and student of Aristotle, first made this leap of faith sometime around 300 BC.

Since then, some of the greatest minds have pondered the possibilities of using tree rings to reconstruct the past. By the time of the Renaissance, Leonardo da Vinci had suggested that there was a relationship between the width of tree rings and water availability, and proposed this could be used to reconstruct past climate. By 1837, Charles Babbage, the 'father of computing', proposed that the patterns of rings in different trees could be overlapped to form continuous records stretching back into the past. By the late 1980s, tree ring experts around the world were doing just that, arguing that a distinctly chilly period started across much of the world in 1628 BC.

Before we get stuck into the 1628 BC event, let's first pull together some of the facts about tree ring dating mentioned in earlier chapters. We can then push onto the boundaries of absolute dating and ask questions not possible with other techniques. Can we get a better understanding of the past by listening to the trees?

The 'father' of tree rings was Andrew Douglass who produced the first dating framework for Arizona in the USA. Trained as an astronomer, Douglass believed that long-term changes in tree ring growth of ponderosa pines were caused by variations in the strength of the Sun over 11-year cycles. Originally, he worked on living trees but in 1914 Douglass started looking further back in time. Archaeologists studying Native American occupation sites, such as at Pueblo Bonito in the Chaco Canyon, and Aztec ones in New Mexico, found ancient timbers in excavations that preserved their tree ring patterns. From these remains, Douglass started overlapping plots of ring widths from individually measured trees to come up with the first continuous 'master' chronology: he coined the term 'dendrochronology' for this method.

Douglass took many years. For some time there was a gap of unknown length between the trees linked to the present day and those that had been found in sites with distinctly different ring patterns. These other trees had to be older, but by how much? Expeditions were organized, using the typological knowledge of Native American pottery to focus on sites where the gap might be bridged. Finally, in 1929, a buried burnt beam was excavated that connected the absolute-dated and the floating chronologies, providing a continuous record of 1000 individual years.

To understand how dendrochronology is such a rigorous dating method, it's worth recapping some basic principles of tree growth. For the purposes of this chapter, we'll restrict our discussion to deciduous trees, and avoid conifers, although the principle is exactly the same. When a tree grows, an annual layer is produced on the inside of the bark, called the cambium. The cambium is made up of two parts. One is the phloem, which transports sugars and other products of photosynthesis through the tree, and later becomes bark. The second part is the xylem; this carries water up from the roots through the trunk and eventually becomes the building blocks of the tree rings.

Xylem cells usually form two parts in a year. The inner or 'early' wood has relatively large cells and forms at the start of the growing season: typically spring, when the factors that control growth, such as nutrients, temperature and moisture, are at an optimum. Later in the growing season, these factors become short in supply, and relatively small cells form with thick walls, much darker than the early wood.

Over the longer term, conditions can change, producing narrow and wide rings, depending on whether the climate and environment were good or bad for growth during that year. Douglass famously could recognize the pattern of rings in many sites just by looking at freshly excavated wood samples. Often, the pattern of thin and thick rings was so distinctive, he could give the age to the year just from memory. Douglass's approach is the basis on which tree ring chronologies have been built around the world: cross-sections of trees are mounted, polished and the ring widths measured to compare and overlap with other samples.

The important point to take home is that because each ring has to overlap with another from a different tree, the method give dates with zero age uncertainty. No other dating method is this precise.

Sampling is not always easy and in extreme cases can even be damaging to your career. In one of the best-known examples, a young scientist, who shall remain nameless, was doing tree ring work on a living stand of bristlecone pines in 1964. Getting his corer stuck in what looked like a stunted tree, he reported the problem to a ranger who offered to fell it for him so he could retrieve his device. When the hapless researcher counted the rings of the felled tree, he realized it had 4950. The tree was growing during the time of the construction of the Great Pyramid of Khufu on the Giza plateau. He had killed the oldest living organism on the planet for a tool that he could have replaced for a day's wages. The young man never worked as a dendrochronologist again.

An excellent example of where dendrochronology has been used successfully (and not to the detriment of a career) is in Denmark. In a sea fjord at Roskilde, five Viking ships were found between 1957 and 1959. The preservation was excellent, largely because of the low oxygen levels at the bottom of the fjord. It appears the ships were scuttled by local people to defend their settlement against potential attacks by other Viking groups. The question was, when were the ships sunk? Although an age for the felling of the trees would not tell archaeologists when this happened, it would give them a maximum date.

All but one of the ships were tree ring dated against local chronologies to the end of the tenth century AD. One ship, however, just would not cross-date. The pattern of the rings did not fit any of the master chronologies from the region. Someone suggested that this particular boat had design features similar to those associated with British and Irish Viking settlements. Some wood was sent to Queen's University Belfast to compare against the Irish tree rings.

Their suspicions were confirmed. It seems that the ship was built in the ancient Viking city of Dublin, with timbers from trees that had died in AD 1042. Intriguingly, when King Harold lost at the Battle of Hastings in AD 1066, some of the remaining Anglo-Saxon royal family, including Harold's queen and son, fled to Ireland and from there to Scandinavia. Could it be that the ship that took them out of the British Isles, when all was lost, has been found, some 900 years later?

☒

In 1999, Mike Baillie at Queen's University Belfast suggested a radical interpretation of global tree ring datasets. Looking at chronologies from around the world, Baillie reported at least four sustained environmental downturns that each appeared to last around four to five years. Unusually, these four took place at exactly the same time around the world. One of these events was originally associated with Santorini: 1628 BC (Figure 6.1). Baillie now recognized others during 2345 BC, 1159 BC and AD 536, with more possible events in 207 BC and 44 BC. These would not have been pleasant times to have lived. Four or five poor growing seasons would have led to successive years of failed crops and been enough to put many societies in serious danger of collapse. Indeed, if this were to happen again, it would put even today's technological society at severe risk.

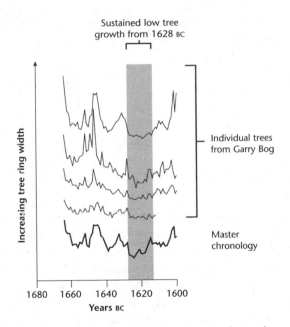

Figure 6.1 Oak ring patterns for trees growing during the 1628 BC event at Garry Bog, Northern Ireland

But it's hard to imagine a mechanism that would account for the several years of poor growth. The identical ages witnessed across the world indicate that whatever took place must have had a global impact. Because the 1628 BC slump had been linked to Santorini, it was originally thought that the events might have been due to major volcanic eruptions.

But volcanic eruptions are now not considered to have had the catastrophic global effects that were envisaged a few years ago. Certainly, super-volcanoes, such as the Yellowstone Caldera in the USA, have had a major impact on life. Most eruptions, even of the magnitude of Santorini, are unlikely to have led to what the tree rings suggest must have been a global cooling of several degrees centigrade for a number of years. Not only that, apart from 1628 BC, no record of associated volcanic eruptions could be found for the other climatic downturns Baillie identified.

Baillie turned to historical records spanning the dates and reached a rather startling conclusion: comets.

The Earth is bombarded daily by space dust. When it falls through the atmosphere, most of us would recognize it as shooting stars. The key question is to what extent larger objects might reach the Earth. Would they strike the ground or explode in mid-air, releasing a shock wave that would devastate large areas?

To find out what kinds of effect an extraterrestrial impact might have had in the past, we can look at what happened at Tunguska in Siberia. Here, on 30 June 1908, an asteroid around 40 m across blew up in mid-air around 8 km up from the surface. The explosion devastated an area of over 2100 sq km and flattened approximately 80 million trees. There was no impact site. People in Europe reported a very bright night at the time but there was no obvious explanation. The event was only recorded because an intrepid explorer went into the area soon after the airburst and took photos and notes.

For the size of the big chills Baillie has suggested, any extraterrestrial object would have had to have been larger than Tunguska. He considers it likely that comets, made up of rock and ice, caused these events. Asteroids, which are either rock or metallic, are less likely. Comets are what Baillie describes as 'psychopathic ice balls', travelling anywhere between 20 to 50 km per second. Most of those spied from Earth come from the fringes of our solar system, either from the Kuiper Belt, just beyond Neptune, or further out from the Oort Cloud. Occasionally, they are knocked out of these areas, into an orbit that may fall on a collision course with Earth. Fortunately, most comets are drawn to the biggest planet in the solar system, Jupiter: its large gravitational field effectively acting as a shield. In 1994, for instance, Jupiter's southern hemisphere was hit by the largest cometary impact ever to be predicted and observed by scientists. Over 20 fragments from the comet Shoemaker/Levy 9 struck. One piece was only 3 km across but hit Jupiter with a force of 6 million megatons, 600 times more than the entire arsenal of weapons on Earth.

But a global chill doesn't necessarily need a direct strike. When a comet orbits past the Sun, part of the ice and dust vaporizes, forming a cloud behind it: a 'tail'. Recent work shows that comets are made up of much more rock and dust than ice. The dust from a large enough tail could fall into the atmosphere, reflecting the incoming Sun's rays, and cooling the planet. This sort of interaction could conceivably lead to crop failures and result in famines, disease and ultimately societal collapse. In such an event, there would be no impact crater.

⊠

Comets are often culturally associated with catastrophes and famine. Baillie cites many biblical references. For instance, the Angel of the Lord often has a bright halo and flaming garment. Similarly, a serpent or dragon could be interpreted as a representation of a fireball leaving a trail through the heavens, known as a 'bolide'. Indeed a nineteenth-century Hebrew encyclopedia describes a comet: 'because of its tail, is called kokbade-shabbit (rod star)'. Intriguingly, Moses is reported to have thrown his rod to the ground to become a serpent.

Babylonian sources record comets in the twelfth century BC, including the note that 'a comet that rivalled the sun in brightness' was observed. Meanwhile, in Ireland, the god Lug is associated with the slaying of a dragon. His name derives from the Celtic for 'light'. Lug was youthful and glorious: his face could not be looked at directly because it was so bright. Were these all representations of comets?

Chinese historical accounts are among the most detailed records available, comparable to those from Egypt, although the dating is often less precise. These archives indicate that environmental events often gave rise to the doctrine of the 'Mandate of Heaven'. If an emperor did not rule his people wisely, Heaven withdrew its blessing, the failed emperor was deposed and the Mandate passed to someone else. So when the sky darkened, the crops failed and the resulting famine led to death, the emperor was blamed and the Mandate of Heaven was believed to have been withdrawn. The result: the replacement of one dynasty with another.

When we look at China, an intriguing pattern of dates can be seen. The end of the Xia dynasty was around 1628 BC, while the close of the Shang happened about 1159 BC. More-over, the boundary between the Ch'in and Han dynasties seemingly occurred in 207 BC. All these are apparently the same sort of time as those events identified by Baillie. There is even a record of what happened to King Jie, the last Xia king. Historical documents from the time refer to the fact that heavy rain toppled buildings and 'The earth emitted yellow fog … the sun was dimmed … three suns appeared … frosts in July … the five cereals withered … therefore famine occurred …'. Could the Mandate of the Heavens have been driven by cometary interaction with the Earth, Baillie asks?

The events have similar environmental changes associated

with them, spanning several years, in different parts of the world. In addition to the USA and Northern Ireland, narrow tree rings are also recorded in England and Germany in 1628 BC. Meanwhile, the Old Testament refers to the exodus of the Israelites at around this time, following dust, ashes and darkness falling on Egypt, cattle being killed by hail, water poisoned and fish dying, culminating in the parting of the sea.

The event of 1159 BC was the worst one recorded in the Irish tree rings. Unfortunately, the broader effects of this are harder to tie down chronologically, compared to 1628 BC. There are few bristlecone pines spanning this period, while the end of the Shang dynasty is not precisely dated. Meanwhile, the Irish king-lists record a 'catastrophe' sometime between 1180 and 1031 'BC', although this must be treated cautiously as dating records from this source is a bit woolly. 1153 BC is also the conventional date of the Egyptian famine. Intriguingly, the number of years between the events of 1628 and 1159 BC is 469 years. Two different Greenland ice cores also give spacings of 479 and 477 years between non-volcanic acid peaks, suggesting they're recording the same events as the trees.

The most recent of the events Baillie identified occurred around AD 540. This is soon after the events surrounding King Arthur. Virtually all trees in Europe show a distinctly chilly time between AD 536 and 545. Pines from Scandinavia record AD 536 as the second coldest summer over the past 1500 years, and that the cooling from AD 541 continued up to AD 550. Overall in Europe, the downturn seems to have taken place in AD 536, followed by a brief recovery, and then with significant cooling stretching from AD 540 until AD 545. The same trend is preserved in the bristlecone and foxtail pines of the USA. Even Douglass referred to the AD 536 ring in the southwest USA as 'often microscopic and sometimes absent', while similar shifts are recorded in South America. The Irish Annals make specific references to a 'failure of bread' in AD 536 and 539 and this appears to be the same in timing as famines in China.

During this period of calamity, the Justinian plague came out of Egypt in AD 542, striking down around a third of the European population. Zacharias of Mithylene even reports that in the eleventh year of the reign of Justinian (AD 538–539), 'a great and terrible comet appeared in the sky for 100 days'. In Gibbon's *The History of the Decline and Fall of the Roman Empire*, eight years after the AD 530 Halley's Comet visit,

> another comet appeared to follow in the Sagittary; the size was gradually increasing; the head was in the east, the tail in the west, and it remained visible about forty days ... The nations who gazed with astonishment, expected wars and calamities from their baleful influence; and these expectations were abundantly fulfilled.

Meanwhile, the 'densest and most persistent dry fog on record' was recorded between AD 536 and 537 in the Mediterranean, while Michael the Syrian observed that:

> the sun became dark and its darkness lasted for 18 months. Each day it shone for about four hours, and still this light was only a feeble shadow ... the fruits did not ripen and the wine tasted like sour grapes.

In Chinese records, several references are given to dragons in the sky. Although there are often problems with dating documentary sources, it is intriguing that all these recorded events are around the same date. Could these passages all be referring to the effects of a brief encounter with or impact by a comet?

In 2004, astronomers Emma Rigby, Mel Symonds and Derek Ward-Thompson of Cardiff University investigated a cometary impact for AD 540. They worked out that a comet just 300 m across could cause the effects implied by the historical observations and trees. Using the results obtained from Shoemaker/Levy 9, the authors suggested that as the comet plunged through the atmosphere, it would have left a hollow tube behind it, into which the surrounding air would not initially have had time to rush back. Like the tube of a gun barrel, much of the energy of an airburst would have been focused back up into the atmosphere with much of the comet debris: perfect for lighting up the night sky and spreading the comet's dust into the atmosphere.

An airburst would have produced enough energy to generate forest fires but due to the height of the explosion would have been too far up to have blown any of them out. Gildas, the depressed sixth-century British monk (Chapter 2), referred to widespread fires and destruction at this time, but most have assumed it was part of his tirade against everyone else. His compatriot, Roger of Wendover, however, who was based in St Albans, observed in AD 541 that 'a comet in Gaul, so vast that the whole sky seemed on fire. In the same year there dropped real blood from the cloud … and a dreadful mortality ensued.' Perhaps Gildas had every reason to be upset.

Chapter 7

THE COMING OF THE ICE

The silent touches of time
EDMUND BURKE (1729–1797)

Imagine a landscape dominated by ice and snow, shaped by long winters, blustery winds and sub-freezing temperatures. This glacial vision is so ingrained that it might seem to have been an idea kicking around for millennia. Yet, just a few hundred years ago in Western Europe, most people believed the world was only around 6000 years old. The rocks, the earth and all the fossils they saw covering the landscape were from the catastrophic Great Flood described in Genesis. It was gospel. But hardly anyone seriously believes this any longer. So what changed? Why is it that we're so comfortable with the idea of ice ages? And can they give us any insight into what the future might hold?

As recently as the late eighteenth century, people saw natural devastation littering the landscape of Europe. It seemed to be everywhere. Even up in the mountains there were jumbles of boulders. What else could have caused such 'catastrophism' but the Great Flood? Yet in 1787 a Swiss minister called Bernhard Kuhn dared to think differently. He courageously suggested that surface rocks and boulders found in areas with a different geology were transported there by glaciers. Now known as 'erratics', these were clear proof to believers of a catastrophic flood. To Kuhn, it was evidence for a natural process.

At around the same time, Scottish geologist James Hutton, one of the founding fathers of geology, began to argue that given enough time, processes seen today could lead to the long-term formation of mountains. He believed that later erosion would form the sediments that filled the bottom of the lakes and seas. We now call this process 'uniformitarianism', although in fairness Hutton shouldn't be blamed for the choice of word: it wasn't conceived until after his death.

In 1795, he developed his ideas in a carefully argued two-volume book called *Theory of the Earth*. The book was almost as famous for its dreadful prose as it was for its science. Being virtually unreadable, it led his friend John Playfair to remark that 'the great size of the book, and the obscurity which may be justly be objected to many parts of it, have probably prevented it from being received as it deserves'. Those who managed to work their way through this tome would have read Hutton's claims that glaciers may have transported the erratics found in the Jura Mountains. Hutton's arguments were a direct challenge to catastrophism. Natural processes, like glacial advances, could explain the modern world. You didn't need to invoke a series of catastrophes.

Despite this, not many made the call to arms. The idea just floundered. It was only in the early nineteenth century that things really started to happen, thanks to one solitary Swiss mountaineer. Jean-Pierre Perraudin lived all his life in the Swiss Alps and regularly saw rocks that had been gouged out. Perraudin believed this must have been caused by glaciers moving over the surface in areas that were now ice-free. Unlike Kuhn, however, he managed to generate enough interest in the idea so that it gathered a momentum of its own. After much perseverance, he managed to convince two engineers to present the concept to the Swiss Society of Natural Sciences in 1829 and 1834. A member of the audience listened to the arguments and was so irritated by the claims that he decided he would prove it false once and for all. His name was Louis Agassiz. At only 25 years of age, he was a rising star in Swiss academic circles.

But things didn't go the way Agassiz planned. By 1836, fieldwork in the mountains had forced him to do a complete

U-turn. He was now thoroughly convinced of the forces of past glaciers. In the following year, he held a position of power that allowed him to speak on it with some authority: he was the president of the Swiss Society of Natural Sciences. At its annual meeting, he was due to give a lecture on fossil fishes, something on which he was an acknowledged expert. Instead, he changed the lecture to the 'ice age', using the term for the first time in an academic meeting. Agassiz was so keen on the concept that he dragged members of the audience up into the mountains to see the evidence. This included grooves etched across the surface of rocks that he believed had formed when stones frozen into the base of a glacier had been dragged over the landscape. The hardline members of this learned society remained unconvinced. After all, they retorted, a horse-drawn carriage could have formed these grooves.

His enthusiasm remained undiminished and, in 1840, Agassiz wrote a book on the ice age that took the evidence to the extreme. In it, he claimed life was wiped out by one rapid major expansion of ice sheets – the 'Great Ice Age'. In the same year, he gave talks in Britain on the subject. Happily he talked about mammals being frozen 'at the time of their destruction'. Despite his extreme views and the suggestions that his ideas were catastrophist, he had soon convinced some of the most influential British geologists of the time, including William Buckland and Charles Lyell, both of whom we shall return to later (Chapter 10). Bestselling author of *Principles of Geology*, Lyell was a fierce supporter of Hutton's uniformitarianism, and was not initially convinced by the catastrophist-inspired ice age that Agassiz seemed to endorse. His old mentor Buckland convinced him otherwise. In late 1840, a united front was presented. Agassiz gave a talk at the Geological Society of London, with supporting lectures from Buckland and Lyell on the topic. The ice age had arrived.

∑

Once the mid-nineteenth century scientists had accepted the idea of ice ages, they raised a critical question: what actually caused them?

Before we start to discuss the causes, it's worth revisiting the basic principles of how the Earth orbits the Sun, some of which we briefly explored with the dating of the pyramids (Chapter 4). We'll start by looking at what happens over the course of a year and then see what changes take place through millennia.

If you've ever bought a globe, you'll have noticed it sits at an angle on its pedestal. Of all the different aspects of the Earth's orbit, the angle to the Sun is the one that drives the seasons as we know them. It was also the first facet of the Earth's orbit to be discovered: by the Alexandrian astronomer Eratosthenes, who lived between 276 and 194 BC. At the moment, the angle is set at 23.5° from the vertical. The effect of this can be seen during the northern hemisphere summer solstice around June 21: the northern half of our planet points towards the Sun and receives the maximum amount of heat possible (Figure 7.1). Six months later, the exact opposite happens. The northern hemisphere points away from the Sun around December 21 on the winter solstice; due to its orientation, the heat from the Sun drops away to a minimum. The important point is that it's not the distance to the Sun that causes the seasons, but the direction our planet is facing. At the moment, the northern hemisphere summer actually occurs when the Earth is furthest in its orbit from the Sun.

In 1605, German astronomer Johannes Kepler recognized how the Earth orbits the Sun. He realized that the planets, including the Earth, didn't travel around the Sun in a perfect circular motion as first thought, but followed an elliptical orbit. Before this time, there had been a long-puzzling observation that days during one half of the year were marginally longer than the rest. Kepler worked out that an imaginary line joining the Sun to a planet sweeps over an area of space in a fixed amount of time. The Sun is actually slightly off centre, so that when we're closer to our star in half of the year, we have a greater angle to it and therefore travel faster. Although the European calendar made no adjustment for this, astronomers in parts of India had spotted the difference. Impressively, they developed a calendar that reflected the different length months: those that took place when the Earth was closest to the Sun had fewer days, while those furthest away had more.

It was realized early on that to understand ice ages, these different controls on the Earth's orbit might hold the key. In 1842, French mathematician Joseph Adhémar made the earliest stab at this in a book called *Revolutions of the Sea*. In it, Adhémar suggested that there had been lots of ice ages in the past and that the cause of these was the shape of the Earth's orbit and the precession of the equinoxes (see the dating of the pyramids in Chapter 4). Because of the shape of the Earth's orbit around an off-centred Sun, the northern hemisphere currently spends several days longer in the 'summer' phase than its winter. The consequence of all this, Adhémar reasoned, was that Antarctica gets more dark, winter nights – it must be gradually getting colder as it receives less heat each year.

To get an ice age, Adhémar suggested that the precession of the equinoxes was the crucial factor. As we learnt earlier, this process changes the orientation of the planet, affecting the relative position of the seasons in the Earth's orbit around the Sun over a 26,000-year period. Adhémar knew that the northern hemisphere summer was currently at its furthest point from the Sun. As a result, in 13,000 years, the opposite would be the case. He argued that because of this, ice ages must happen in the hemisphere where winter is at the furthest point from the Sun. The ice ages occur in the hemispheres at different times.

It was a brave attempt but hopelessly wrong. By 1852, it had been shown that precession did not vary the amount of heat from the Sun – the solar radiation or insolation – over a year. Both hemispheres got exactly the same amount of heat throughout the year. This couldn't drive an ice age. But he was right about one thing. During the 1860s and 70s, geologists started finding plant fragments between leftover glacial landforms in Scotland and North America, showing there had been more than one 'Great Ice Age'. Adhémar had planted the seed of an idea. Could there be an alternative?

The challenge was taken up by the British scientist James Croll. Croll was an amazing man. He'd lived several lives, as a wheelwright, a tea merchant and then a temperance hotel manager, before he became a janitor at the Glasgow Andersonian College and Museum in 1859 at the age of 38 years. Croll desperately wanted to have access to its library. By 1864, the janitor had published his first paper on multiple ice ages. He argued that changes in the shape of the Earth's orbit from elliptical to nearly circular and back to elliptical (its 'eccentricity') over 100,000 years had a major role to play. But, unlike Adhémar, Croll was not concerned with how much heat the Earth received over a year.

Instead, Croll argued that it is the way the heat is distributed through the year that matters. In a highly elliptical orbit around the Sun, the Earth receives more heat in one season than another. When the planet is furthest from the Sun, it has exceptionally chilly winters. Croll argued that major snowfields would build up if you had a number of successive cold winters in a highly elliptical orbit. Because of the reflectivity, or albedo, of the snow, he reasoned, the growing snowfields would increasingly reflect what little radiation was making it to the planet's surface. It would get even colder – a positive feedback. Precession only played a role in this when the eccentricity was high. When it was, Croll agreed with Adhémar that the ice ages must happen in the hemispheres at different times.

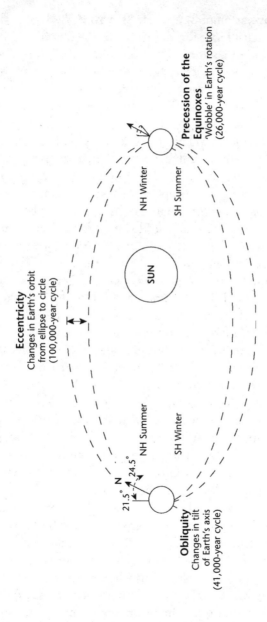

Figure 7.1 The different controls on the Earth's orbit around the Sun

As the Earth was known to be travelling around the Sun in an orbit that was more circular in shape, Croll argued that it didn't matter what precession was doing. The orbit was not sufficiently elliptical to build up enough ice for an ice age.

But Croll was not finished. In 1875, he introduced into the equation the third and final astronomical feature of the Earth's rotation: the tilt or 'obliquity' of the planet. By the turn of the nineteenth century, it was known that the planet could actually nod backwards and forwards between 21.5° and 24.5°. Croll suggested that when our planet is tilted at a greater angle, ice ages are less likely as the poles would receive more heat through the year. All these factors indicated to Croll that there couldn't have been an ice age for at least 80,000 years. Since then, it had been relatively warm – a period known as an 'interglacial'.

What was needed was an independent date for the last ice age. Remember this was well before the time of radiocarbon dating, which did not arrive on the scene until the mid-twentieth century. To try to get a handle on the past, some researchers were using sedimentation and erosion rates to calculate how long features such as lake deltas and waterfalls would have taken to form since the ice melted. Estimates were all over the place and had enormous uncertainties but clustered between 10,000 and 20,000 years ago. Could these ages be believed? If so, it would be a significant blow to the orbital theory.

In the late nineteenth century, it was discovered that many lakes fed by melting glaciers often filled with a distinctive pattern of sediments. Glaciers are rarely pure ice. Instead they often contain large amounts of different sized mineral grains that they pick up and crush when travelling over the landscape. During the spring and summer, some of the ice melts and, rich with sediment, flows into adjoining lakes. At these times, the heavy sandy particles settle rapidly to form a layer on the lake floor. Over the year, as the melting dwindles, the finer, lighter particles left in the water finally settle on top of the layer of sand.

At this time, Swedish scientist Gerard de Geer discovered these layers in old lake deposits within areas that used to be covered by ice. He likened the regular combined layers of coarse and fine sediments to tree rings and reasoned they must represent individual years. De Geer coined the term 'varve' and argued that counting them would reveal how many years a glacier had fed into a lake. Because varves are related to the amount of ice that has melted, the thickness of an individual layer can vary year to year, from just millimetres to several centimetres in thickness. Glaciers in an area would have responded to the same climate and should have produced similar patterns of varve thickness in adjacent lakes. Using the same principle as tree ring dating, these patterns could be compared and overlapped.

From 1878, armies of students were taken out into the Swedish countryside by de Geer to compare varves from lakes originally formed next to retreating ice at the end of the last ice age. The lakes had since dried out and, for-tunately for de Geer, streams and rivers had cut through the bottom, exposing the layers. By 1910, he could clearly show that there had been an enormous icecap over the whole of Scandinavia. But the timing was all wrong. The start of the ice retreat was around 10,000 years ago, and not 80,000 years as suggested by Croll – this was a major problem for the orbital theory.

It was really one man, a Serbian called Milutin Milankovitch, who spent a good part of the First World War reworking Croll's ideas, who finally cracked the case. In 1920, Milankovitch had calculated the combined influences of eccentricity (the 100,000-year cycle), obliquity (the 41,000-year cycle) and the precession of the equinoxes (the 26,000-year cycle) on the amount of solar heat received at different latitudes for the past million years. Milankovitch argued that it was the land at high latitude, in particular 65°N, that was the key: this was where the largest changes in the amount of solar heat took place.

The major leap made by Milankovitch was that he argued it was low summer warmth that allowed snow to survive through the year that was important. Only when the maximum temperatures stayed low could ice persist and build up. This was the opposite to Adhémar and Croll, who had both insisted that extremely cold winters were needed to start an ice age. His results were startling. Instead of predicting the ice age to have ended 80,000 years ago, Milankovitch's new interpretation indicated that it finished around 10,000 years ago, similar to the sediment evidence from de Geer and others.

This seemed to fit in nicely with the end of the last ice age, but what about earlier? If there had been more ice ages, did the orbital theory explain these as well? The problem was there was no way of testing it on the land. The last ice age had destroyed almost all the landforms created by earlier ones. Just small pockets of evidence survived here and there. What was needed was one long record stretching back in time that showed what the ice was doing.

The answer came from an unexpected direction.

⌛

Let's recap on what we have so far. At the end of the seventeenth century, people had started spotting strangely grooved rocks in mountainous areas of Europe, many of them with a geology that was different to the area they were found in. At the time, most people felt this was consistent with the Great Flood described in Genesis. By 1840, Agassiz had decided that all these were actually the result of a Great Ice Age. Between the 1860s and 1910, Agassiz was vindicated, but there were now known to have been multiple ice ages in the past, the most recent one only ending around 10,000 years ago. What had caused them was not known, but, by the 1920s, Milankovitch had shown that changes in the way the Earth orbited the Sun over thousands of years was a good bet. The only problem was that no one really knew when the earlier ice ages had taken place.

Up until now all the fuss had been on land. Few people had bothered with the oceans. It was only from the 1930s that scientists had started to drive long metal tubes from research ships into the ocean floor – coring – and analyse the sediment they collected. Most believed the ocean environment hadn't changed much in the past.

This all changed in 1955, when an Italian, Cesare Emiliani, decided to look at the shells of foraminifera preserved in long cores from the ocean that spanned hundreds of thousands of years. Affectionately known as 'forams', these small creatures live at different depths in the ocean water, and when they die, their shells often become buried in the sea mud. Emiliani believed that stable isotopes preserved in the forams might hold the key to understanding what the climate had been in the past.

Isotopes, remember, are atoms that have the same number of protons but a different number of neutrons. Although we have mostly looked at combinations of protons and neutrons that are radioactive, most are actually stable. The result of all this is that once fixed by an organism, the ratio of one stable isotope to another stays the same from day one. No matter how much time has passed, the signature of the stable isotopes should stay the same.

Emiliani was interested in reconstructing temperature using two stable isotopes of oxygen called ^{16}O and ^{18}O. Try to visualize them as two balls of different weights. ^{18}O being heavier than ^{16}O because it has two more neutrons. The important point here is that chemically they behave the same way.

The beauty of using forams is that they take oxygen directly from the ocean water to build their shells of calcium carbonate. It's known from analysing modern forams that as the water gets colder, they fix more of the heavier balls of oxygen; a shift often described as going 'positive'. When it gets warmer, more of the lighter balls of oxygen are anchored instead; the forams become more 'negative'. When Emiliani looked at the ratio of the different oxygens in the ancient forams down his cores, he was stunned: there seemed to be a cycle of warm and cold temperatures over the past 300,000 years. The shape of the temperature curve was similar to that predicted by the orbital theory for ice ages. Could this be the proof that Milankovitch was right?

But there was a potential hitch. Did the isotopic signal in the forams really record temperature? Modern studies had shown that forams did just that, but what happens if you go back to an ice age. Were the rules of the game the same?

In an ice age, not only is it cold but as a result there's a lot less evaporation from the ocean's surface. Over time, the heavier water molecules tend to stay in the sea water; it's far easier for what little evaporation takes place to remove water made up of the lighter balls of oxygen. At high latitudes, this evaporated water condenses and falls as snow, forming vast ice sheets. In other words, the ^{16}O is preferentially taken out of the ocean and locked up in the ice, while the sea becomes richer in ^{18}O. But when an interglacial takes place, the opposite happens: more of the heavier balls of oxygen evaporate as water under the warmer conditions. Meanwhile the ice melts, returning the locked-up ^{16}O to the ocean. The result: the ocean has relatively less ^{18}O in it. So ice volume could be a major control on the oxygen isotopic signature in the forams over the long term.

In the 1960s, the American John Imbrie and British Nick Shackleton suggested that sampling anywhere too close to the poles would feel the combined effects of changes in temperature and ice volume. Ironically, if you wanted to get the best snapshots of past ice, they argued you had to look at the tropical ocean sediments. The ocean acts as one enormous conveyor belt, taking warm surface water into the North Atlantic – popularly known as the Gulf Stream – and returning south as cold, dense water at the bottom. This deep ocean water finally comes back to the surface through upwelling several centuries later, forming the final phase of the cycle. Because of this, the ocean is well mixed; when the ice melts at the poles, the change in oxygen isotope content of the water is rapidly transmitted around the world and taken up by the forams when they form their shells. Since the temperature in the tropics would have changed a lot less in the past, forams from here give a purer record of changing ice volume.

The fact is that Emiliani's temperature interpretation on the oxygen isotope curve was just part of the story. The problem was that most ocean sediments built up far too slowly to precisely test the orbital predictions of 100,000-, 41,000- and 26,000-year ice cycles.

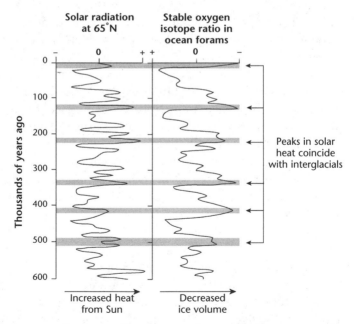

Figure 7.2 Changing ice volume and solar radiation for the past 600,000 years

In the mid-1970s, attention was fixed on two ocean cores from the Indian Ocean. Changes in the Earth's magnetic field and radiocarbon-dated forams in the core showed they had an unusually high rate of sedimentation. This meant the cores could be analysed at far closer time intervals than had been possible before. Could these be the proper test of the orbital theory? The forams were extracted and measured for their oxygen isotopes. The scientific community was on tenterhooks. The result: the changes in ice volume showed exactly the same cycles as the orbital theory predicted (Figure 7.2). The cycle times of eccentricity, obliquity and precession of the equinoxes were all there. Here at last was direct evidence that the changes in the Earth's orbit around the Sun controlled the ice ages. Adhémar, Croll and Milankovitch had been right all along.

8

In order to understand future climatic change, we need to be able to study periods of rapid shifts that happened in the past. Unfortunately, ocean records rarely record rapid climatic changes, and where they do, precise dating is difficult. So researchers started scouring other parts of the world for sites with long, detailed records of climate change. Attention soon turned towards the polar icecaps.

At the poles, snow that falls each year is preserved as ice going back many thousands of years. Buried deep and trapped over millennia are a whole host of different features of the climate and environment: dust, acidity, volcanic ash, green-house gases and isotopes. In Antarctica, climatic changes spanning around 800,000 years have been reconstructed. The 100,000-year cycles predicted by the orbital theory can be seen as clear as day. In Greenland, the record only goes back undisturbed 123,000 years, but each individual year can be counted. The results are beautifully detailed climate reconstructions from these regions; something rarely possible from the oceans.

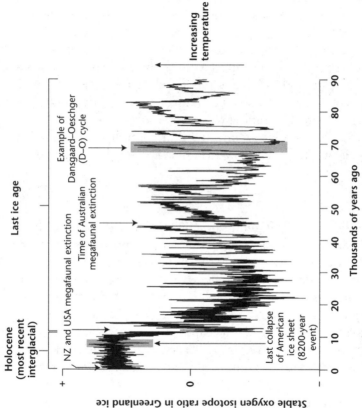

Figure 7.3 Temperature changes in Greenland over the past 90,000 years

Note: The timing of megafaunal extinction is discussed in the next chapter

Frighteningly, the records from Greenland ice show massive and frequent shifts in temperature between 90,000 and 11,550 years ago. Called Dansgaard–Oeschger events (Figure 7.3), the temperature swings are almost as much as when the climate goes from an ice age to an interglacial, but all within just a few years. Nothing like this is predicted by the changes in the Earth's orbit around the Sun. So what's going on?

A clue might be in the ice-core levels preserved around 8200 years ago. Here a 200-year cold snap marks the melting of the last remaining body of North American ice, a tiny remnant of the last ice age. All the resulting freshwater hurtled into the North Atlantic, capping the surface of the ocean and effectively stopping cold, dense sea water from forming. This deep water formation is the driver of the ocean conveyor belt we briefly mentioned earlier. 8200 years ago it looks like it got a near-fatal hit, almost shutting down completely. The Gulf Stream that brings warmer water north was seriously disrupted and high latitude areas got much colder. It was almost a mini-ice age in the north.

If this is the cause of the Dansgaard–Oeschger cycles, it looks like our world is quite happy switching rapidly from warm to cold and back again far more frequently than we'd like to think. An extreme of this idea formed the basis for the 2004 movie *The Day After Tomorrow*. Although fairly far-fetched, if the ocean is more sensitive than we thought, any future melting of polar ice could shut down the North Atlantic conveyor belt pretty much instantly, with dire consequences for the north and perhaps globally.

Chapter 8

THE LOST WORLDS

Time, the devourer of everything
OVID (43 BC–AD 17)

There was a time – yesterday, geologically speaking – when kangaroos 3 m tall hopped about Australia, elephants roamed North America and 2-m high birds strutted around in New Zealand. Since the end of the nineteenth century, many different writers have noticed that the world does not have abundant creatures over 40 kg. Collectively, these large beasts are known as 'megafauna'. Alfred Wallace, who wrote the first paper on evolution by natural selection with Charles Darwin, noted that 'we live in a zoologically impoverished world, from which all the hugest, and fiercest, and strangest forms have recently disappeared'. What happened to them all and are we to blame?

We now know that the extinction of many of these creatures was global and that they died out quite recently. Their bones, when explorers and researchers found them, were not fossilized, suggesting their death was a matter of thousands of years ago. But the extinction seems to have happened at different times in different places. Some regions even kept their megafauna. In Australia, 94% became extinct, while at the other extreme, only 2% were lost south of the Sahara. What happened?

As with any good mystery, there are two main suspects: in this case, climate and humans.

The idea that our animal-skin clad ancestors may have hunted the huge beasts to extinction was first suggested as long ago as the mid-nineteenth century. Several major criticisms continue to be levelled at this theory. One is that many large animals are still present in Africa, despite it having the longest record of occupation by people (more of which in the next chapter). Also, to its detractors, the fact that there were not huge numbers of our ancestors at the time of the main extinctions suggests they could not have caused large amounts of environmental damage. And it has also been controversially argued that most animals are shy of humans and are unlikely to hang around long enough to feel the hard end of a club.

The alternative is that a rapidly changing climate caused the habitat of the megafauna to shrink or disappear. This seems pretty attractive as an idea. As we saw earlier, at the end of the last ice age, there was a major change in the global climate. By around 10,000 years ago, things had started to warm up so that the climate was almost comparable to today. Animals that were adapted to icy conditions, the argument goes, were unlikely to be able to cope with a rapid transition to a warmer climate. A major criticism here is that there have been other major climatic changes in the past, some of which have been equally extreme and rapid. What could have been so different at some climatic boundaries to have caused widespread extinction when earlier shifts had had no discernible effect?

※

An excellent test of some of these ideas is what happened in Australia. As well as the giant kangaroos, a large number of different species are now sadly extinct. One of the best known is the giant herbivorous marsupial, the diprotodon. This creature was furry and wombat-like. Up to 2 m high and around 3.5 m long, it would have looked more at home in a *Star Wars* movie. When you throw in the now extinct marsupial lions, sheep-sized echidnas (spiny anteaters) and large goanna-like carnivores (monitor lizards) over 5.5 m long, you have to wonder what happened.

The problem with the Australian fossils is that they often appear to have lain on the surface for some time, losing most of their carbon content before finally being laid down in the sediments for archaeologists to find them. Because of this, the bone itself is often useless for direct radiocarbon dating, while the sediments they're buried in are often an entirely different age.

Is there another way of dating the megafauna in Australia? One of the best examples where a different angle was tried was with the dating of the extinction of the largest of the Australian flightless birds. Weighing up to 200 kg and standing 2.2 m tall, *Genyornis newtoni* was common across much of central and southern Australia. The few skeletal remains there are show its legs were quite short and thick, suggesting it was a slow runner. Fortunately, it was a prolific layer of eggs. Over large parts of Australia, distinctively smooth eggshell pieces are often found poking out of sand dunes. These eggshells have recently been dated using a number of methods by Giff Miller, from the University of Boulder, and colleagues.

The eggshells are made out of calcium carbonate, so can be radiocarbon dated. With this method, Miller's team soon found that all the ages came out around 40,000 years old. If you remember from Chapter 3, this age is suspiciously close to the limit of the method for many laboratories. After several half-lives of 5730 years, not much original radiocarbon remains in a sample. Other dating methods had to be tried to date the extinction of *Genyornis*. Two different types of approaches were brought to bear: 'amino acid racemization' and 'luminescence'.

Amino acid racemization exploits the way that the organic chemicals in shell, bone and wood change over time. The first work in this area was done in the 1950s and the principle is relatively simple. Eggshell, while mostly calcium carbonate, also contains proteins, which are made up of building blocks called amino acids. Amino acids can come in left- and right-handed varieties, chemically identical but structurally mirror images of one another. After an animal or plant dies, some of its amino acids flip into their mirror image variant. The practical upshot of this is that when you analyse a modern eggshell, the amino acids will all be the left-handed type. Over time, though, these molecules begin to convert to a right-handed mode. Older samples have a greater proportion of right-handed amino acids.

Although around half of all amino acids do decompose over time, enough material can survive to allow researchers to measure the ratio of the mirror images of certain amino acids. This ratio is a guide to the length of time since the organism died. The beauty of this method is that the preparation is relatively quick and cheap, allowing literally hundreds of analyses. The problem with amino acid racemization is that it just gives a relative age. Another dating method is needed to calibrate the amino acid ratio to a calendar age. Radiocarbon would be ideal but *Genyornis newtoni* appeared to be too old. So Miller and his colleagues dated the sand where the eggshells were found, largely using a method called luminescence.

Luminescence dating is relatively new. In contrast to radiocarbon, it has the advantage of working on inorganic matter and can date back at least 800,000 years. The technique calculates the time since mineral grains were last exposed to light or heat. One of the great disadvantages is that anyone who works with this method has to spend most of their time toiling in the dark, with nothing but a faint red torch for company.

The principle of the method is based on the fact that when a mineral, such as quartz or feldspar, is formed, it does not have a perfect structure. Over time, any radioactive isotopes in the ground will undergo decay and nearby buried minerals feel the effects of this: the energy produced excites some of their electrons, knocking them out of their orbits. In most cases, they return to where they came from, giving off a tiny photon of light. But sometimes, the imperfections in the mineral's structure trap them. Crudely, we can think of the defects in the mineral as traps gradually filling up over time with excited electrons. When a sample is exposed to sunlight or heat, these trapped electrons get enough energy to escape and return to the atoms they had left.

The more electrons trapped, the longer the sample has been buried. To calculate how many electrons have been trapped, samples have to be brought back to the lab in a black plastic bag or tube to stop sunlight resetting the clock. In the darkened lab, with only the trusty red torch, the samples are either heated ('thermoluminescence') or exposed to a particular wavelength of light ('optically stimulated luminescence') to get the trapped electrons to escape. When this happens, the amount of light given off can be measured. Meanwhile, the sediment surrounding a sample is analysed for its radioactive content to calculate how much energy the mineral grains experienced from decay in the ground. As shell fragments don't last long on the surface, the enclosing sand grains must have been last exposed to sunlight at about the time the eggs were laid. The important point is that by working out how many electrons have been trapped and at what rate they were being collected by the mineral grain, an age can be calculated.

Miller applied these methods to similarly sized emu eggshells. He found them to span the last 120,000 years, right up to present day. Importantly, there was no age bias in the dataset. They'd got samples evenly distributed throughout this window of time. When the same dating exercise was done on the *Genyornis newtoni* eggshells, the result was significantly different. The last individuals of this species appeared to have lived around 50,000 years ago, far beyond the 40,000 year limit of radiocarbon. It was the first rigorous attempt to get a handle on the extinction of huge creatures in Australia. The question was: how representative was the bird of the other megafauna?

Tim Flannery at the South Australia Museum and Bert Roberts at the University of Wollongong in Australia had a crack at this question. Unlike Miller, they looked at bone remains across much of the country, but to avoid potential problems, they looked at sites where the remains were still joined together – 'articulated'. If the bones were in a jumble when they were excavated, the animal could not have died where it was found, and the enclosing sediments might not represent when the creature died. Using optically stimulated luminescence of sediments surrounding articulated remains, Roberts, Flannery and colleagues found the megafauna had died out across Australia approximately 46,000 years ago. Although the mean age was 4000 years younger than the results from Miller's team for *Genyornis newtoni*, the time of extinction overlapped within the dating uncertainties.

The similarity in the dates suggests the same cause for their extinction but what was it? One method used to investigate this was to look at the eggshells and identify what sorts of plants *Genyornis newtoni* lived on. As we saw with dating the Turin Shroud (Chapter 3), in addition to radiocarbon, carbon also has two stable forms: ^{12}C and ^{13}C. Plants have different stable carbon isotopic contents: wet-lovers tend to comprise more ^{12}C, while those that thrive in dry conditions, particularly many grasses, have relatively more ^{13}C. By measuring these different isotopes, it's possible to see what was delectable to *Genyornis*.

The results were intriguing. Apparently, *Genyornis newtoni* lived almost exclusively on wet-loving plants, while the emu, which has happily carried on foraging through to today, appears to eat both wet- and dry-loving plants – it was more omnivorous. We mentioned in Chapter 7 that this was a time of extreme climate change in the North Atlantic, but as far as the Australian region goes, not a lot appears to have happened. If diet was important but the climate wasn't doing much, was the isotopic data a red herring?

Could humans be the key missing element? To explore this idea, we have to consider when people arrived in Australia: a debate that has raged for over 40 years.

Over the course of the last ice age, the amount of water locked up in the vast ice sheets around the world would have resulted in a drop of sea level of up to 130 m. Although this sounds huge, a fall in sea level of this magnitude would have only led to the joining up of Papua New Guinea, Australia and Tasmania. This great landmass would have remained an island detached from Asia. Early populations who reached Australia must therefore have built a craft that could cope with an ocean crossing.

In the early 1960s, it was thought that humans had only accomplished this sometime within the last 10,000 years. Since then, the date has been systematically pushed back. By 1995, radiocarbon ages of between 38,000 and 40,000 years for arrival were known from archaeological sites in Western Australia. This all seemed perfectly reasonable. Most of the archaeological community seemed content.

But as we said before, 40,000 years is suspiciously close to the conventional limits of radiocarbon dating. In particular, the late, great archaeologist Rhys Jones suspected this age was a result of the dating method and not real. He and Bert Roberts looked at sites in Arnhem Land, in Northern Territory, where there was no charcoal but the great depth of artefacts suggested early arrival. In 1990, using luminescence on grains from the deepest levels of artefacts, they announced that Arnhem Land was occupied between 50,000 and 60,000 years ago. You could have heard a pin drop.

⌛

It does not take much modern carbon to shift a radiocarbon age when little or no original ^{14}C remains. One per cent contamination can give a sample an apparent age of 37,000 years, even if it was really formed millions of years ago. Although charcoal is often found in archaeological sites as a result of human activity, these small amounts of contamination can become quite an issue if most of the initial radioactive carbon has decayed away. Just remember what shoe insoles to remove foot odour are made of: charcoal. It soaks up almost anything. It's indiscriminate. Put it in the ground for tens of thousands of years with rainwater percolating through the sediments, and you have a potentially major problem on your hands if you want to date beyond 40,000 years. While I was at the Australian National University, I worked with Michael Bird and Keith Fifield on a new method of cleaning up charcoal called ABOX; short for Acid-Base-Wet Oxidation. We found ABOX removed most carbon contamination that other methods couldn't reach. The method produced clean charcoal that could be radiocarbon dated back to 60,000 years ago.

Using ABOX and other techniques, we studied a small limestone cave site called Devil's Lair in Western Australia. Here, the original radiocarbon dates couldn't get beyond 40,000 years. Was this real or a fault in the method as Rhys Jones had suggested? A team which also included Mike Smith at the National Museum of Australia and Charlie Dortch at the Western Australian Museum, took charcoal samples through the levels with the deepest artefacts. The results fulfilled our wildest dreams. We found that humans had lived in the area of Devil's Lair around 48,000 years ago. The ages broke right through the radiocarbon barrier. We'd cracked it. This was the earliest radiocarbon age for human arrival in Australia and supported the luminescence dates from Arnhem Land. Humans really had reached the continent at around the same time as the extinctions had taken place.

But no remains have been found in Australia that unambiguously show people were hunting and butchering these huge animals. So can humans still be blamed? It might be that we just haven't found a site yet. After all, 46,000 years is a long time ago. Yet perhaps there's a reason why no kill sites have been found. One further clue may lurk in the stable carbon isotopic ratios of the *Genyornis newtoni* eggshells. They suggest some sort of environmental change took place at the time of the extinction. Could fire have driven this change?

It has long been known that Australian aborigines made widespread use of fire for hunting and removing pests. James Cook, for instance, referred to Australia as 'This continent of smoke' when he sailed past it in 1770. Lynch's Crater in northeast Queensland – an infilled extinct volcano – contains sediment recording environmental change spanning at least the last 200,000 years. With Peter Kershaw at Monash University, a group of us analysed the uppermost sediment layers for different pollen types to see what the vegetation was doing in the past. Alongside this, we also measured the amount of charcoal preserved in the sediment layers as an indicator of burning.

There was a sudden and dramatic increase in the amount of fire 11 m down. This depth records a time when there was a dramatic long-term shift away from rainforest plants to dry and fire-tolerant vegetation, such as eucalypts. Nothing like this had ever happened during the previous ice ages around Lynch's Crater. It had to be people. Radiocarbon dating suggested that the burning started 46,000 years ago – statistically indistinguishable from the megafauna extinction. Perhaps by burning the vegetation when they arrived, people altered the Australian environment so much that the landscape could no longer support larger animals. If so, could this explain extinctions elsewhere?

⧗

North America lost slightly fewer of its huge beasts than Australia – around 73%. They were in many ways equally bizarre: the giant sloth, which stood 3 m tall and weighed around 2,500 kg; at least two species of horse; a camel; the mastodon, which was related to the mammoths and modern elephants; and the Colombian mammoth that reached up to 3.4 m. Here, remains have largely been dated using radio-carbon because, in contrast to Australia, the extinctions appear to have happened far more recently.

In North America, radiocarbon dating seems to show the majority of extinct megafauna disappeared as recently as 11,400 years ago. Intriguingly, the mastodon and mammoths managed to survive a little later until 10,900 years ago, suggesting there could have been two separate phases of extinction. Either way, these dramatic events are a full 35,000 years later than their equivalents in Australia. Why were global extinctions happening at different times in different places?

There's no doubt that this was a time of upheaval. Major climatic and environmental changes were underway in North America at the time of the extinctions. It was the beginning of the long and painful recovery from the last ice age. We know the ice started to retreat from around 17,000 years ago. Moreover, charcoal and pollen preserved in lake sediments across North America show that from around 15,000 years ago, the temperature rose enough to allow the development of closed forests but with no significant burning. This seems to have bought about the disappearance of the steppe grassland that may have evolved alongside the megafauna through the previous ice age. If fire wasn't the cause, could rapid warming have led to a nutritional bottleneck for the animals? Stable carbon isotopes in mastodon and mammoth remains certainly suggest so. These species seem to have had specialized diets, similar to *Genyornis newtoni*, making them vulnerable to big environmental shifts. If they couldn't adapt quickly enough to the changing conditions, they were in serious trouble.

Complementing all this, scientists have managed to extract genetic material in the form of deoxyribonucleic acid – better known as DNA – from soil and lake sediments in the North Pacific region. This gives a fascinating insight into what the environment was like at the time. The DNA comes from excrement left by roaming beasts and reveals a rapid drop in herb vegetation and an increase in mosses at the same time as the extinction of the megafauna. On these results, the case for climate change wiping out the North American animals seems more convincing than for Australia.

Many researchers think not, blaming humans instead. In North America, both sides of the debate are more equally matched than in Australia. Extinction seems to have coincided with significant climate change and human arrival.

Unlike Australia, North America has regularly been connected to Asia via the Bering Strait. During drops in sea level over successive ice ages, the Bering Strait became a land bridge between the two continents. We know modern humans were in northeastern Asia around 30,000 years ago, but this was also a time of high sea levels. The traditional overland route through the Bering Strait into North America would have been submerged at this time. Using radiocarbon, we know for sure that human populations reached what is now Alaska sometime around 13,000 years ago; possibly when it was warm enough at the end of the ice age to explore north of Siberia and cross the Strait, but while sea levels remained low. These early groups then seem to have stopped short. Although the ice had started melting 17,000 years ago, most of Canada and much of the northern USA remained covered. The traditional view was that only when enough ice had melted to open up a corridor could people gain an entry point into the interior of the continent.

The first clear evidence for human arrival in the interior of North America seems to have been the Clovis people. They are named after the small town in eastern New Mexico where, in the 1930s, their distinctive fluted stone spearheads were found alongside mammoth remains. Radiocarbon dating now shows they were established around 11,300 years ago.

If the Clovis did come through an opening in the melting ice after crossing Beringia, you would expect to find similar styles of tools in Siberia. As we saw with the Scandinavian Bronze Age, typology is a useful method for plotting migrations of people and ideas. The problem is that no Clovis-like tools have been found in the Siberian region, even though this is felt to be their most likely origin. It's almost as if the technology appeared from nowhere. Bizarrely, the closest Clovis-style artefacts actually come from 16,000–19,000-year-old sites in southern France, associated with the Solutrean people. The idea that these people could have crossed the North Atlantic by canoe, hugging the ice, is heresy to many archaeologists. Time will tell.

As if things weren't complicated enough, we now know humans had reached South America before Clovis. This is despite the enormous distances involved. At Quebrada Jaguay in southern Peru, remains of fishing excursions made by humans have been radiocarbon dated to 11,100 years ago, while at Monte Verde in central southern Chile, evidence of hunter-gatherers has been radiocarbon dated to 12,500 years ago. None of the tools look anything like the Clovis ones. This all suggests that the peopling of the Americas was a lot different to that originally envisaged; some of our fur-clothed ancestors apparently didn't need a land corridor.

These dates suggest that there was a lot more going on at the end of the last ice age than originally thought. Rather than one route of entry, there may have been multiple migrations of people from many directions, using both land and sea to settle the Americas. Border control would have been a nightmare.

Where does all this leave us with megafaunal extinction? Although there might have been migrations of other people into the Americas, North America has clear signs that the Clovis people hunted the large beasts. There are at least 12 sites, including the original Clovis find. Most spectacularly, in Naco, Arizona, an adult mammoth was found with eight Clovis-made spear points embedded in the skeleton.

Dating neighbouring island animals in the region allows an

excellent test of the relative importance of climate and humans in driving megafaunal extinction. On St Paul Island in the Bering Sea, radiocarbon-dated mammoth remains show that after getting trapped by rising sea levels at the end of the ice age, they lounged around until at least 7900 years ago. In Cuba, radiocarbon dating of the now extinct ground sloth showed it was happy eating leaves until at least 4960 years ago, similar in time to when humans reached this island. If climatic and environmental changes were the main cause of extinction in North America, it's hard to explain how these groups carried on happily while their cousins on the mainland disappeared several thousand years earlier.

To test some of these ideas, computers have been increasingly used to model past extinctions. The arrival of a human population in these studies leads to a surprisingly rapid rate of animal disappearance. In many cases, extinction happens during simulations representing only a couple of hundred years. It doesn't seem to depend on the size of the species or the land area involved: those greatest at risk of extinction were groups with low reproduction rates. Slowly reproducing populations didn't stand a chance against any form of hunting. They rapidly became extinct. These models suggest that you wouldn't expect to find the last individuals of an extinct species associated with a stone tool. As long as enough of them had been taken out of the gene pool, the rest of the population would struggle on until the demise of their kind.

☒

Refreshingly, New Zealand gives us an absolute answer to the cause of the mass extinction of its large creatures, which included the giant Haast eagle. Probably the best known were the 11 species of moas. The largest of these flightless birds were over 2 m tall and weighed up to 250 kg. Moas seemed to breeze through all the climate changes during the ice age that might have caused so many problems elsewhere. Despite the proximity to Australia, and their similar climatic influences, the moas happily carried on strutting about the New Zealand landscape 46,000 years ago. In contrast, the megafauna across the Tasman Sea were dropping like flies. What happened? People.

New Zealand was the last significant landmass in the world to be colonized. Radiocarbon dating of rat remains suggests that humans may have arrived as early as 2200 years ago. Because the Pacific rat is not native to New Zealand, it must have arrived there with the early colonizers, probably as a food source. The problem is that there is no direct archaeological evidence for human occupation. No settlements or artefacts date back to this time. Nothing. And exhaustive studies of pollen in sediments show no characteristic markers of human activity in the landscape, such as significant burning or vegetation change. If humans did indeed arrive when the earliest rat ages suggest, there can't have been significant settlement. It certainly didn't lead to the widespread colonization of New Zealand. The moa carried on regardless through this time.

The first clear archaeological evidence for human settlement is a lot more recent: radiocarbon dating puts it at around 700 years ago. Settlements, widespread forest clearing by fire and stone tools are found in abundance across much of the country at this time. Coincident with all this, the moa seems to have rapidly become the choice cuisine. At some sites, it almost seems that people didn't live on anything else. Dismembered moa remains dominate sites. In fact, they're so common in archaeological contexts that they're often used as a sign of early human occupation: the 'Moa-hunter' period.

Estimates of how long moa hunting went on suggests the birds' extinction may have taken just a few hundred years. By 500 years ago, moas were scarce. The rapid collapse of moa in each area probably took less than 20 years. The end of the main course in the area forced people to move into previously unexploited areas. The most recent moa remains indicate that the animals were probably extinct by 1700. By the time the first Europeans landed in New Zealand in the eighteenth century, there were no living moas to be found anywhere.

⊠

So where does all this leave us? Shifts in climate must have had an impact on past populations of colossal animals but it's hard to see this delivering the knockout blow. Those species now extinct had all evolved alongside earlier changes that were often just as rapid and extreme. Perhaps most telling of all is that in the cases we've looked at, the dates of human arrival are close in timing to the extinction events. It's not a hard call to believe that when humans arrived in virgin lands, it was just too much change for many of the megafauna. Perhaps they were already weakened by a changing climate but those creatures that couldn't avoid our ancestors were up against it from the start. They never really stood a chance.

Chapter 9

AND THEN THERE WAS ONE

A very merry, dancing, drinking,
Laughing, quaffing, and unthinking time
JOHN DRYDEN (1631–1700)

The 'missing link'. The phrase encapsulates an array of ideas: a creature somewhere between apes and ourselves; eccentric scientists exploring the back of beyond; a deep-seated desire to know where we came from. The expression often guarantees media time for any find, no matter how small. Yet it's also one of the most divisive areas of human endeavour. With only a small number of human fossils, there are almost as many ideas about what they mean. You can almost guarantee sparks will fly when a new find is announced, often before the ink has dried.

In truth, the 'missing link' is a dreadfully out-of-date concept. The term was coined soon after the publication of Charles Darwin's *Origin of Species* in 1859. In it, Darwin was able to show that only evolution could explain the vast range of species found today and in the past. He reasoned that on the basis of their similar biology and behaviour, humans were most closely related to chimpanzees and gorillas. The missing link was the hypothetical species between the apes and ourselves. As we'll see shortly, we now know from fossil evidence that there wasn't just one species making the link. There were lots of them.

Before we get buried in what human remains have been found and their age, it's first worth briefly looking at how individual finds survive the upheavals of the past to come to us today as fossils. In a general sense, the term 'fossil' is often used to mean any cast, mould or impression of a once living thing. It's not very precise; just a useful turn of phrase to tell fellow workers that we're not looking at a modern sample. In its strict sense, however, fossil refers to mineralized bones, shells and plant material.

Within just a few weeks of death, scavengers and bacteria usually start to break down and recycle all body tissues. To avoid this and end up with a mineralized fossil, the tissues have to be rapidly buried in sediments. There's a number of ways this can happen: a volcanic eruption; a freak flooding event; an earthquake; taken as prey to the lair of an animal – anything that leads to part or all of a body being rapidly concealed from normal decay processes. Given enough time, water may then leach the minerals from bones and allow alternative ones to be deposited, turning the remains to stone. Even then, their survival is not guaranteed. Geological processes may destroy what has endured up to that point. In many respects, it's amazing that anything survives through the ages to reach us today.

⧗

Although nature seems desperate to cover her tracks, there are enough fossil human remains to give us some strong clues to what happened in the past, much of it in Africa. We know that sometime between 5 and 7 million years ago, humans and apes went their own separate ways. We also know that by at least 4 million years ago, our ancestors were walking upright in the form of *Australopithecus* – or 'southern ape'. By 2.5 million years ago, our genus *Homo* turned up on the African scene in the form of *Homo habilis*, with a greatly enlarged brain compared to *Australopithecus*, and wielding stone tools.

Back in the nineteenth century, none of this was known. When Darwin's ideas were published, the race was on to find the missing link. Not everyone believed Darwin was right that the origins of humankind lay in Africa. Based on the observations that gibbons were able to walk upright and that they lived as nuclear family units, German biologist Ernst Haeckel suggested a different location: Southeast Asia.

Haeckel's ideas had a major impact on our understanding of human evolution, but not in the way he originally envisaged. In the late 1880s, a visionary Dutch researcher called Eugène Dubois took up Haeckel's challenge. Finding he could not get research support to pursue his ideas, he applied for a position as a medical officer in the Dutch army and left his promising academic position in the Netherlands to take his family to Indonesia in late 1887.

Starting in Sumatra, he convinced the local authorities that he should be released from his medical duties to pursue his research. In 1890, he relocated to Java due to the better preservation of fossils there. Initially, Dubois had focused his efforts on cave deposits but found the amount of fossils to be disappointingly low. Switching to low-lying areas, Dubois concentrated on where the rivers were cutting away old terraces. These had accumulated over time as sediments washed down the valleys into the river systems, becoming rich in fossil remains.

In 1893, Dubois coordinated a dig on the Solo River near the settlement of Trinil in central-east Java. The site has not changed much since Dubois' time. Dense forest comes down to the river and the excavation pits dug a century ago in the remaining terrace are still visible. Even today you can see why Dubois chose this bend on the river: ancient fossils of now-extinct animals are still found sticking out of the sediments. Working away at the site with two engineers and a crew of labourers, he found a skullcap, a thighbone and a tooth from within the same tranche of sediment. The skullcap was clearly different to a modern human. It was thicker, with a brain size intermediate between apes and our own species, *Homo sapiens*. Dubois had got what he came for.

Packing up his finds and expecting high praise, Dubois

returned to Europe. He was disappointed to find many of his peers sceptical. Rather than being greeted as a hero, many chose to ignore his find. As we'll see later with other ancient human finds, the response ranged from the sublime to the ridiculous. Some supported Dubois' claim that it was the crucial part of the fossil puzzle, but others suggested that he'd got it wrong and the fossils were mixed-up remains of different species, or they represented an ape, or the old favourite, a sick modern human. Fortunately, in the late 1920s and 30s, identical finds were made by Ralph von Koenigswald at Sangiran in central Java – 'Java Man' – and Davidson Black in China – 'Peking Man'. The new species became known as *Homo erectus*. Dubois, who lived until 1940, was vindicated. Ironically, he considered these other finds to be of some other intermediate form between humans and apes, absolute in his belief that the Trinil find was the true missing link.

Since Dubois, *Homo erectus* has now been found throughout Asia and Africa. But when was it kicking around? Rarely can we use radiocarbon for ancient human fossil remains. Although the samples would originally have contained radiocarbon, they are generally much older than the 60,000-year cutoff point; all the ^{14}C has long since decayed away. Thankfully, many of the earliest finds in Indonesia and Africa have been found near centres of volcanic activity. Many different volcanic strata have been found, often covering vast areas and encasing skeletal remains. While we can't date the bones directly, we can determine the ages of associated volcanic layers.

One of the earliest methods applied to dating early human remains was argon-potassium. The element potassium, which has the chemical symbol of 'K', has three forms. The version we're interested in is ^{40}K because it's radioactive. As with all radioactive isotopes, ^{40}K decays; it has a half-life of 1250 million years. The reason it's so popular for dating early human sites is that potassium is extremely common in different volcanic rocks. Sometimes, when radioactive potassium decays, a proton in the nucleus captures an electron turning it into a neutron and forming an isotope of stable argon gas, written as ^{40}Ar.

The important point is that when ash and rock from a volcanic eruption cools and hardens, there should be no argon gas present. But when the potassium within the rock begins to decay, argon gas starts to form and becomes trapped. If the gas within the rock can be extracted, the amount of ^{40}Ar that has built up can be measured and an age calculated for the volcanic eruption and any associated skeletons. This dating method was proposed in 1948 and first applied in 1965 to some of the first *Australopithecus* finds, doubling the timescale of human origins overnight.

Although this all sounds fairly straightforward, there is a potential problem with potassium-argon dating. Two measurements are needed, one to determine the amount of potassium in a sample and a second to measure the ^{40}Ar. This means a large sample is needed, which opens up a chance for contamination. To bypass this snag, a variation of the method was developed in the 1960s, called argon-argon dating.

Here the sample is irradiated in a nuclear reactor, converting the potassium to another isotope of argon, ^{39}Ar. To get an age we can now just concentrate on the two isotopes of argon gas: ^{40}Ar and ^{39}Ar. By heating the sample, the gas can be captured and the ratio of these two isotopes measured at the same time. The result is that the age for the sample can be determined using a lot less material, cutting the chances of contamination. Heating the sample using either a laser or furnace progressively releases the trapped argon gas towards the centre. This is collected for measurement. If a rock sample is uncontaminated, the ratio between the two different isotopes of argon should always be the same. But if different parts of the sample have not been sealed from the air, atmospheric argon may have leaked in, skewing the age. By heating the sample progressively to higher temperatures and measuring the different argon values through a volcanic rock, a more accurate age can be produced.

At Koobi Fora, in Kenya, argon-argon dating of volcanic material surrounding *Homo erectus* skeletal remains and stone tools gave an age of 1.88 million years old. But the best-known example of dating *Homo erectus*, however, is 'Turkana Boy': a spectacular, 90% complete skeleton of a 10–12-year-old, discovered by Kenyan archaeologist and anthropologist Robert Leakey on the eastern shore region of Lake Turkana. The argon-potassium dating gave an age of 1.64 million years for the volcanic matter couching this specimen.

So how old were the earliest *Homo erectus* in Java? A date for the find at Trinil has still not been done, largely because no one has found suitable volcanic grains for argon-argon dating. But von Koenigswald found other examples of *Homo erectus* across Java. Without any method of directly dating his finds, von Koenigswald looked at the location of the sites in the landscape where *Homo erectus* had been found and the different animal remains preserved in the sediments; these suggested to him that one particular site called Mojokerto was older than Trinil. The American Carl Swisher, from the Berkeley Geochronology Group, and his colleagues undertook argon-argon dating of a volcanic layer believed to be associated with where the Mojokerto skull was found. This suggested *Homo erectus* may have been in Java as early as 1.81 million years ago. But the precise location of where von Koenigswald made the find is still hotly debated and recent work puts the find down to as recent as 1.43 million years. It looks like *Homo erectus* was on the other side of the world after they'd first evolved in Africa.

⌛

So it seems that sometime between 1.8 to 1.4 million years ago, *Homo erectus* picked up its tools and decided to move out of Africa towards Indonesia. Why? No one knows for sure, but it now seems likely there wasn't just one migration out of Africa. Several *Homo erectus*-like skulls have now been found in Georgia, dated to 1.8 million years. And by around 800,000 years ago, an offshoot of *Homo erectus* had turned up in Europe, eventually evolving into arguably one of the best-known extinct human species: the Neanderthals.

Homo neanderthalensis was actually the first ancient human species to be recognized in the fossil record. The first find was made in Gibraltar in 1848 but was largely ignored. In 1856, a more complete skeleton was found in a German limestone quarry in the Neander Valley. This made people sit up and take notice. The discovery was made three years before the publication of *The Origin of Species* and it caused quite a bit of consternation. The original quarrymen who found it had thought it was a bear. An 'expert' argued it was a Mongolian Cossack who had deserted the Russian army chasing Napoleon in 1814. Another stated quite firmly that it was an individual who had rickets as a child, was later knocked on the head and then struggled into old age with arthritis.

Apparently evolving from an offshoot of *Homo erectus* called *Homo heidelbergensis*, early Neanderthal fossil remains are few and far between. The age when we would recognize them as a species in their own right is most definitely vague: sometime between 250,000 and 500,000 years ago. As a species, they had several features quite different to ourselves: stockily built, with pronounced eyebrow ridges, a larger brain case, and no chin. One of the most striking characteristics of a Neanderthal skull is the enormous cavity in the middle of the face, indicating that individuals had particularly large noses. Why they had these features is still open to question, although it has been argued that the large nose was an adaptation to cold climates; warming up frigid air as it was inhaled. Certainly, Neanderthals seem to have dominated high latitudes, when other human species appear to have remained in tropical regions. They inhabited a land that experienced the full expression of interglacials and ice ages. Being geographically isolated in an ever-changing environment, Neanderthals evolved on a different pathway to other species of human.

⊠

Although the early evolution from *Australopithecus* to *Homo erectus* certainly appears to have taken place in Africa, the arguments over what happened to later species continue to rumble on. Where did our own species – *Homo sapiens* – evolve and when? Two competing ideas have come out of the handful of finds.

The 'out of Africa' hypothesis argues that *Homo sapiens* arose in Africa and then moved out into the rest of the world, outcompeting more ancient species. In contrast, the 'multi-regional hypothesis' asserts that species of *Homo* in different parts of the world evolved into *sapiens* in parallel but separate to one another.

Much to the annoyance of supporters of the multi-regional hypothesis, the earliest human remains showing the same features as us are all from Africa: relatively short, flat faced with no pronounced eyebrow ridges and a chin. A virtually complete skull of an early form of *Homo sapiens* has been found in the Middle Awash in Ethiopia and dated to between 154,000 and 160,000 years ago. In 2005, evidence from the Omo River in Ethiopia revealed the earliest known example of our species to be 196,000 years old.

Until recently, it was believed that the Neanderthals and modern humans were blissfully unaware of each other until around 40,000 years ago. The first such contact almost certainly happened in the Middle East. Excavations from the 1920s, particularly in Israel, found a number of key caves where early human remains were discovered. Some, such as Kebara and Amud, contained Neanderthal remains, while other sites, such as Skhul and Qafzeh, had *Homo sapiens*. On the basis of radiocarbon dating, it was believed that the Neanderthals occupied the region around 50,000–60,000 years ago, until *Homo sapiens* moved in 40,000 years ago.

As we've noted several times before, ages of 40,000 years are suspiciously close to the practical limits of radiocarbon dating. Alternative methods were needed to be tried on these Israeli sites. One of these was electron spin resonance dating, often abbreviated to ESR.

ESR works to a similar principle as luminescence dating by measuring the number of trapped electrons. There are several differences though. In most cases, ESR is done on teeth, not mineral grains. Also, instead of releasing the electrons in the lab using heat or light, the sample is put in a changing magnetic field. The more magnetic power the sample absorbs, the more electrons it harbours. The joy of the method is that it can be used to date teeth, regardless of whether they have been sitting in a museum display and exposed to years of light. The electrons analysed by ESR are not in traps sensitive to light.

When thermoluminescence and ESR were applied to the Israeli sites, a quite different story unfolded. The *Homo sapiens* sites of Skhul and Qafzeh were dated to between 90,000 and 130,000 years old, while it emerged that the Neanderthal sites of Kebara and Amud were 50,000–60,000 years old; the opposite to what was expected. The results seemed counterintuitive. If *Homo sapiens* had replaced Neanderthals, how could they have pre-dated them?

The answer almost certainly lies in the climatic changes known to have taken place 90,000–130,000 years ago. In Israel, this last interglacial was potentially too warm for the Neanderthals and probably forced their retreat to cooler, higher latitudes. *Homo sapiens* were better able to exploit the situation and migrated into the region. But 50,000–60,000 years ago, the climate got worse again. The ice age returned. The Neanderthals, adapted to cold conditions, migrated south back into a region they had colonized earlier; *Homo sapiens*, who probably could not cope with the deteriorating conditions, retreated.

It was sometime later that modern humans had another go at moving out of Africa. Some of the earliest evidence for their jumping-off point into the Middle East is from Egypt. Here a child skeleton of *Homo sapiens* has been found. The sediments it was buried in have been dated using luminescence to between 50,000 and 80,000 years ago.

Radiocarbon dating suggests that early *Homo sapiens* swept aside the Neanderthals in the Middle East around 40,000 years ago, and over the course of several thousand years moved across Europe. A massive upheaval of the status quo in both technology and organization took place, with the end result that our species dominated Europe.

The best evidence for the early arrival of *Homo sapiens* in Europe is a modern human skull from Romania, directly radiocarbon dated to 34,000 years old. Radiocarbon dating of Neanderthal bones has shown that they managed to hang on in parts of Croatia until at least 32,000 years ago. Stone tools made by the Neanderthals have been found in pockets of Europe, including southern Spain, Portugal and Gibraltar, and given ages as recent as 30,000 years ago. Intriguingly, a near-complete child skeleton in Portugal appears to show some Neanderthal features. Radiocarbon dating of charcoal associated with this find gives an age of 25,000 years, making this the youngest known Neanderthal individual.

Perhaps our mental capabilities were enough to outcompete the Neanderthals. Some researchers think we may simply have reproduced more frequently and literally outbred the opposition. Alternatively, our better designed tools may have given us the edge, making us more efficient hunters able to beat the Neanderthals to dinner. There is no evidence that the two species fought pitched battles. The Neanderthals were doomed. Genetic studies show there was little if any interbreeding between the species. By 30,000 years ago, they'd been pushed to the fringes of Europe. By 25,000 years ago, they seem to have disappeared off the face of the Earth.

⌛

But what happened to *Homo erectus* in Indonesia? We know they'd arrived by at least 1.43 million years ago. When did they die out? The American Carl Swisher and his colleagues returned to Indonesia and in 1996 reported dating results from a Javan site called Ngandong. Originally reported by von Koenigswald in the early 1930s, eleven skulls were discovered here within one river terrace along the Solo River. The fossils show *Homo erectus* had evolved relatively large brains compared to other ancient Javan finds of this species, suggesting a young age. Swisher's team dated the remains using uranium-series and electron spin resonance.

We've already looked at electron spin resonance but how does uranium-series work? The name alone often has people running for the shelters. Fortunately, the amounts in most natural systems are miniscule. Usually the natural concentrations of this element are of the order of parts per billion, equivalent to one drop of ink in an oil tanker.

While we're alive, our bones contain no uranium. After death and burial, bones act like a sponge, soaking up any uranium dissolved in ground water. The uranium decays to thorium, which is also trapped in the bone. In the lab, these elements can be measured. But it's not immediately obvious how and when uranium was fixed in the bone during its burial and whether any was later lost if the fossil had been exposed at the surface. A mathematical model is needed to work out how the uranium would have migrated into the bone. The age can then be calculated by working out how long it would have taken to get the final measured amounts of uranium and thorium.

Swisher reported ages for the Ngandong *Homo erectus* finds that were totally unexpected. The skulls were as young as

27,000 years. The results sparked a huge debate, not least because they suggested that the species survived for almost a million years longer in Java than Africa. They were potentially alive at the same time as the last of the Neanderthals. Geologically speaking, this was yesterday.

⧖

Against this backdrop, a small island east of Java had been a nagging sore in understanding early human origins. In the mid-twentieth century, Father Verhoven, a Dutch priest and amateur archaeologist, had travelled across Flores, excavating fossil sites. In the Soa Basin of central Flores, he claimed to have found several sites with volcanic sediments containing stone tools, one of which was called Mata Menge. He guessed they were around 750,000 years old because of their association with fossils of stegodon, an extinct species of elephant. At the time, his findings were generally dismissed as wild speculations. He was just an excited amateur.

It might seem odd that the archaeological community was immediately dismissive of Verhoven's claims but there was a reason for this. Indonesia is divided by the biogeographical boundary called the Wallace Line, first identified by the British naturalist Alfred Wallace in the nineteenth century. West of this divide, the flora and fauna are comparable to those of Southeast Asia, while to the east, the living things are more like those in Australia. Changes in sea levels, driven by successive ice ages, frequently connected the western islands to Asia. But even with drops of 130 m or so, those lands to the east remained unconnected and ecologically distinct. The depth of the seabed is too much for a land bridge to link all the islands of Indonesia. As a result, Java and Bali lie west of the Wallace Line; Flores, east.

As far as most archaeologists were concerned, the problem for Verhoven was that 750,000-year-old stone tools in Flores implied that *Homo erectus* must have been capable of breaching the Wallace Line. That just couldn't be right. They hadn't managed ocean crossings anywhere else.

This house of cards started to fall in 1998. Archaeologist Mike Morwood from the University of New England in Australia worked with colleagues on a reanalysis of Verhoven's site Mata Menge on Flores. They dated the finds using a method called fission track.

Fission track dating exploits physical changes caused by the decay of uranium. When this radioactive element disintegrates within volcanic rock or glass, the resulting particles collide with the mineral's structure and damage the grains, forming scars. By counting the physical tracks left on the grains and measuring the concentration of uranium in the sample, an age can be calculated. Morwood found the artefacts from Flores really were stone tools. And the age? 840,000 years. Verhoven had been right all along.

But what had happened to the makers of the stone tools? *Homo erectus* may have got to Java by 1.43 million years ago and survived until 27,000 years ago, but no skeletal remains were known from Flores.

Mike Morwood pulled together an Australian and Indonesian team to continue the work on Flores. The group were interested in finding when *Homo erectus* became extinct in the region and investigating the most probable route for the first modern humans into Australia. Because of my work in Australia using the ABOX method to push back the limits of radiocarbon dating of charcoal, I was invited on board. I didn't realize at the time just how fortunate I was.

The team focused on Liang Bua, a limestone cave in western Flores, near the small town of Ruteng. Verhoven had done some initial excavations in the upper part of the cave in the 1950s. Since then it had been regularly excavated, prodded and raked over, although generally near the surface. The team wanted to go deeper. In the 2003 excavation, most of the group had packed up for the season and the Indonesians, led on-site by Thomas Sutikna of the Indonesian Centre for Archaeology, were polishing off the work for the end of the season.

On 12 September 2003, I got an email from Mike Morwood that was beyond my wildest dreams: they'd found a near-complete human skeleton at a depth of 5.9 m. Alongside were lots of stone tools and evidence that stegodon had been hunted. The worn teeth indicated that the remains were those of an adult but it was only 1 m tall. Apparently charcoal had been found with the remains. Would I be interested in dating some samples? I jumped at the chance.

The cave at Liang Bua is a huge cavern, with massive stalactites hanging from the roof. When I visited the site, the place was a hive of activity. Armies of locals were patiently clearing and sifting the sediment coming out from the excavation. The main trench where the skeleton had been found had got down to around 10 m in depth before digging had stopped. The sides were buttressed with planks of wood and there was a complicated system of ladders and decking for people to get down to different levels. Compared to the stifling heat outside, the whole grotto was a haven of cool, moist air. If you were going to set up camp anywhere on Flores, Liang Bua would be it.

The main find was of an adult female. When she was found, the bones almost seemed to have the texture of blotting paper; the remains hadn't yet turned to stone. She had to be left to air dry for three days before any more excavations could take place. When she was dug out of the ground, it was clear she wasn't a *Homo sapien* or a *Homo erectus*. Many features in the skeleton were unusual. Not only was she short in stature, but her brain was tiny; the cavity where it once was measured a mere 380 cm^3 – similar in size to a chimpanzee. Previously, the smallest *Homo* brain was thought to be around 500 cm^3, and this was for the first known species of our genus, *Homo habilis*, some 2.5 million years ago. The skeleton had many other unusual and ancient features, including a sloping forehead, wide pelvis, arms that reached down to its knees, and teeth made up of many roots. The stone tools suggested these little creatures were not stupid; they could think for themselves.

The find was clearly not related to modern pygmy humans. Despite their small stature, pygmies have similar sized craniums to us and they certainly don't have the other ancient features seen in the Liang Bua human fossil. What could it be? Many of the features of this skeleton seemed more ancient than those of the Javanese *Homo erectus* populations. The little people seemed to have had more in common with the earliest *Homo*.

The charcoal samples arrived soon after I had replied to Mike's email. I didn't dare hope that I would be able to get an age from them. The skeleton traits suggested it was ancient. It should have been off the scale for radiocarbon. Shortly after the samples arrived, I got them prepared as quickly as I dared.

I'll never forget when the results came through. I was at a conference in northern Wales at the time and it was 2 am in the morning. I up against it – I had a presentation to give that morning and was still finishing my slides for the talk. An email came through. I glanced at my inbox to see who it was from. It was Keith Fifield who was running the samples at the Australian National University. I suddenly woke up. The samples from the Liang Bua skeleton were measurable using radiocarbon. I quickly converted the numbers into a calendar age. The results showed it had lived 18,000 years ago. I was dumbfounded and ecstatic! I hardly slept a wink that night.

The implications were enormous. Here was an ancient lineage, apparently derived from one of the earliest migrations out of Africa, which had crossed the Wallace Line, become stranded on an island, evolved and shrunk. Previously it had been thought our own species, *Homo sapiens*, was the only one intelligent enough to make an ocean crossing of several kilometres by raft or log in large enough numbers to establish a self-sustaining population. Yet here was an ancient species that seems to have done it not once but at least three times. Even at times of low sea level during an ice age, the jumping-off point from Bali would have required ocean crossings to the islands of Penida, then Lombok and Sumbawa (which were both joined as one at low sea level), before reaching Flores. There is even the possibility that other islands may have become home to early human populations which then evolved independently into separate species – they needn't have stopped at Flores but could have kept going east.

We announced the find to the world on 28 October 2004 as a new species: *Homo floresiensis*, better known as the 'Hobbit'. Within a few days, the sparks began to fly. Was it really a new species? Could it be a pygmy with a rare disease that led to underdevelopment of the brain? It was a rerun of the debates that surrounded the first discoveries of the Neanderthals and *Homo erectus*. Even more fossil finds of similarly proportioned individuals didn't satisfy the critics. It just showed you couldn't please everybody.

Interestingly, there are several quite detailed folk tales of creatures on Flores similar in description to the shape of the Hobbit and which were documented before the find was announced. Some stories refer to 'Ebu gogo' which means 'ancestor that eats anything'. The name was coined after regular encounters between villagers and these creatures: apparently they would even eat the pumpkin-base plates as well as the food they were offered. Some of these stories are remarkably detailed, suggesting this species may have persisted on Flores until only a few centuries ago.

It is a sobering thought that just 30,000 years ago, up to four species of human might have existed on our planet. Now we believe there is just us. The discovery of a living survivor of another species would really put the cat among the pigeons. Would we shake it by the hand, put it in a zoo or deny that it even exists?

Chapter 10
THE HOLE IN THE GROUND

'Dear me', said Mr Grewgious, peeping in,
'It's like looking down the throat of Old Time.'
CHARLES DICKENS (1812–1870)

Ever since dinosaurs were first identified in the nineteenth century, their disappearance has been a source of fascination. At 65 million years ago, it is the most recent of five mass extinctions in our planet's history – a catastrophic event that wiped out between 45% and 75% of all species living at the time, some of which were arguably the most spectacular organisms our planet has even seen. But how did it happen? How could a world of such diverse life be eradicated in a blink of geological time?

Our knowledge of dinosaurs has built up surprisingly recently. The first dinosaur remains were only discovered in the seventeenth century, mostly within northwest Europe. The earliest description was by the first professor of chemistry at Oxford University, Robert Plot, who, in 1676, described a large bone dug up from an Oxfordshire quarry; although he believed it was most probably an elephant brought to Britain by the Romans. In 1776, a giant crocodile-like skull was discovered in chalk in the Netherlands. The find unsettled the local inhabitants so much that they nicknamed it the 'Beast of Maastricht'.

At first, most of these dinosaur fossils were believed to represent the remains of animals that had been killed in the biblical Great Flood. It was supposed that when the waters of this event had retreated, the fossils were laid down in the sediments created by the devastation. So widespread were these views that a new post of professor of geology was established in 1818 at Oxford University. The first incumbent was Reverend William Buckland, whose remit was to corroborate the biblical Flood.

At the time of these first discoveries, however, geology was just becoming a recognized science. Workers in this new field of endeavour desperately needed a timescale. But without any direct dating, the next best bet was to try and get an idea of the relative age. One of the earliest attempts to do this was developed by a German geologist called Abraham Werner at the turn of the nineteenth century. Werner believed that different rock types could be recognized as one of four kinds that were formed in a strict chronological order.

The first of these types, Werner considered, were the most primitive in the Earth's history and therefore the oldest. These were called the Primary and were made up of schists and granite. Importantly, the Primary rocks contained no fossils so were believed to be pre-Flood. Immediately above the Primary were the Transition rocks, which included limestone and slate, and these contained a small number of fossil remains. Above them were the Secondary rocks, which were often layered and included limestone and sandstone. For believers in the Flood, this type was the most important: they were packed with fossils, supposedly from the catastrophe. Finally, the uppermost type was the Tertiary, which was represented by loosely bound rock types that included clay, sand and gravel.

Geology almost immediately struggled to reconcile the biblical accounts of creation with fossils recorded in the rocks. The sheer scale of the sequences identified by Werner and seen across much of Europe implied a considerably longer period of time than the 6000 years suggested by theologians. Geology was on an early collision course with the Bible.

⋈

Of all the early fossil hunters, probably the best known was Mary Anning, who made a living from the things she discovered within the Secondary coastal cliffs of Lyme Regis in Dorset, England. Mary Anning became something of a national celebrity and is believed to have inspired the tongue-twister 'She sells seashells on the seashore'. A poor family, the Annings collected fossils to supplement their income. With the death of her father Richard in 1810, Mary took up the pursuit in earnest and in 1811 and 1812 she and her brother Joseph discovered the first remains of an ichthyosaur. Popularly known as the 'fish lizard', this appears to have been one of the first reptiles to become fully adapted to life in the sea.

Eleven years later in 1823, Anning went one further and discovered the first almost complete skeleton of a 3 m long creature. With a small head, four flipper-like fins and a long neck equivalent to the length of its body, she'd found what we now know as a plesiosaur. It was a monster. Even today, it forms the basis of one of the more popular 'reconstructions' of the Loch Ness Monster. But at the time it was new to science and totally unexpected. The neck was felt to be almost too long to be real.

From then on, the pace of discovery and description of dinosaur fossil remains shot up. In 1822, the British country doctor and geologist Gideon Mantell made the first scientific description of dinosaur bones extracted from rocks in Sussex, England. He likened them to giant lizards. On 20 February 1824, Mary Anning's plesiosaur remains were fully described at the Geological Society of London. At that same meeting, William Buckland suggested that equivalent gigantic beasts also lived on land, when he described the remains of a meat-eating reptile, megalosaur – the earliest of the giant bipedals. This was followed up in 1825 by Mantell who described to the Royal Society the remains of a large, slow-moving plant eater called an iguanodon, which he suggested was a reptile.

In the early nineteenth century, these fossil finds raised

major problems for a strictly literal interpretation of the biblical version of the origins of the world. Nowhere in the Bible was there reference to a prehistory and prehistoric animals. Yet in the fossil record, they were found in abundance. When the text was analysed closely, other details also failed to hold up: insects were reputedly created after mammals, directly contradicted by the geological record.

At least one extinction was now being recognized by geologists. How this was explained depended on whether you were a 'catastrophist' or a 'uniformitarian'. The great French scientist Georges Cuvier suggested that the geological record preserved local extinctions. In contrast, William Buckland favoured one global extinction, caused by the Great Flood. Alternatively, Louis Agassiz proposed that the Great Ice Age had been the cause; something that Buckland agreed with after the Swiss scientist had visited Britain in 1840 (Chapter 7). Meanwhile, Charles Lyell believed that extinctions were natural events that happened predictably – they were entirely consistent with uniformitarianism.

With the recognition of past extinctions, came another dangerous idea. If species could die out, then they were not permanent. All of life could not have been created in one single act. This implied there had been a progression over time. An advancement towards more complex life. This was a tempting idea.

Although there was increasing support amongst the scientific community for evolutionary processes to explain the geological record, it was not one-way traffic. Some dissented. Probably the best known of these was the British biologist Richard Owen. He tried to preserve the argument for an ordained reason for the existence of species. An excellent anatomist, Owen was the first to describe many of the extinct megafauna from Australia and New Zealand, including the diprotodon and the moa. Unfortunately, he also had a tendency to try to claim credit for work done by others. In one incident in 1844, he presented results to the Geological Society on belemnites; marine, meat-eating, squid-like creatures that lived around the same time as dinosaurs. Identical work had been expounded to the same august body just a few years earlier by another scientist. As if this wasn't enough, wherever possible, Owen seemed to go out of his way to belittle Mantell's efforts and downplay the significance of his competitor's dinosaur finds.

Despite these major character flaws, he correctly realized in 1842 that the finds made and described by Mantell and Buckland of the iguanodon and megalosaur were different from living reptiles. Owen argued that these remains were not earlier ancestors in an evolutionary chain but represented entirely different creatures. He proposed a new name for the extinct group of reptiles: 'dinosaurs', derived from the Greek *deinos* meaning 'fearfully great', and *sauros* meaning lizard; sometimes simplified to 'terrible lizard'. Perversely, this had the opposite effect to what Owen intended. With an ever-increasing number of fossil finds, his identification of dinosaurs as a separate subgroup was later used to great effect by evolutionists, including Darwin, to demonstrate a series of progressions in life, from the simple woodlouse-like trilobites in Transition rocks to mammals in the Tertiary.

⊠

With no direct dating and only a simple geological scheme for different rock types, confusion reigned. We now know that some of these early workers were mixing up extinct creatures of different ages, including dinosaurs and megafauna. Of the five mass extinctions encapsulated in the geological record, the end of the dinosaurs was not the largest, just the most visible to the pioneering researchers. Earlier extinctions had taken place 200, 251, 375 and 444 million years ago. The dubious honour of the largest event goes to the Permian extinction – 'The Great Death' – at around 251 million years ago; up to 95% of all species became extinct. Interestingly, the disappearance of the megafauna we looked at in Chapter 8 doesn't even count as a mass extinction when it's put in a geological context.

The 'time of the dinosaurs' is now known to fall within the Mesozoic era. This consists of three periods: the Triassic, 200–251 million years ago; the Jurassic, 146–200 million years ago; and the Cretaceous, 65–146 million years ago. The rise of the dinosaurs took place at the end of the Triassic, probably following the mass extinction of other species around 200 million years ago. Throughout the Jurassic period, larger versions began to evolve, so that by the Cretaceous, the greatest diversity of dinosaurs existed. But why, at their peak, did these magnificent creatures disappear?

The end of the dinosaurs is often referred to as the 'K-T' event or boundary. The 'K' comes from the German word for chalk, *kreide*, marking the Cretaceous period; the 'T' comes from the 'Tertiary' originally proposed in the scheme by Werner. The geological stratigraphical framework has been revised since the K-T was originally adopted. Strictly speaking, the 'T' should be dropped and replaced by 'P' for Palaeogene, but the acronym has stuck.

Some Cretaceous limestones originally laid down in the deep sea are now exposed on land in Italy, Denmark, New Zealand and the USA. Visiting one of these sites is an extremely humbling experience. You can go right back in time, geologically speaking, to the end of an era. One excellent example of a K-T site is Woodside Creek in New Zealand. This is just a 20-minute walk into the hills from the main road. Awaiting you is a cliff of late Cretaceous and early Neogene limestone, tilted at a slight angle and cut through by an eroding stream. Near the base of this cliff, the cream-coloured limestone of 'K' is topped off by a layer of dark clay, just 1 cm thick. It's close enough to the ground to comfortably put your finger on the exact spot when the whole world changed. Atop this thin layer is a darker limestone, marking the beginning of the Palaeogene period, representing the start of 'T'.

Marine fossils are commonly found in the lower creamy-coloured limestone of the Cretaceous period, and are sometimes large enough to be seen by the naked eye. But above the clay layer in the darker limestone of the Palaeogene, there are virtually no fossils and those that have been found are microscopic. At all the sites spanning this period, the same sequence of sediments is seen. In other words, the K-T boundary was a global event. But what could have caused it?

By the 1960s, several origins had been suggested for the K-T boundary: climatic change, volcanic activity, and one or more meteorite impacts. The only way to find the cause was to precisely date the evidence of all these. This would show whether any coincided with the K-T boundary. The chosen method was potassium-argon and argon-argon dating that we looked at with dating human origins (Chapter 9).

One strongly supported suggestion for dinosaur extinction was that a series of volcanic eruptions took place at the K-T boundary. One of the best contenders for this was the Deccan Traps of India. These represent the largest known phase of volcanic activity from the time. During the Cretaceous period, the distribution of continents on the Earth's surface was significantly different to today. At this time it appears that India was migrating north towards Asia, over a hot spot that currently sits under Reunion Island in the Indian Ocean. The resulting volcanic eruptions produced lots of lava layers that formed an enormous plateau. Also known as 'flood basalts', these traps cover an area the size of France – at least 500,000 sq km – and represent approximately 1 million km^3 of lava.

The Deccan Trap eruptions must have gone on a long time to produce all this lava. They may have pumped vast amounts of ash and gases into the atmosphere, potentially shielding the Earth from the Sun's rays, cooling the surface. Such changes would have significantly reduced photosynthesis and driven extremely rapid climatic change, potentially leading to global extinction. Dates put on the Deccan Traps in the 1960s and 70s supported this model. Ages of between 40 and 100 million years ago were obtained using potassium-argon and argon-argon dating. Unfortunately, these ages were not precise enough to say whether the volcanic activity led, lagged or peaked with the K-T boundary known to be around 65 million years ago.

At the same time as these Deccan Traps studies, an alternative hypothesis was being developed. In 1980, a team led by the father and son duo of Luis and Walter Alvarez from the University of California, tried measuring a range of different elements in the thin dark K-T clay layer. They were interested in those that are abundant in meteorites but rare in the crust and upper mantle of the Earth – elements such as iridium. When meteorites burn up on entering the Earth's atmosphere, the iridium and other elements they contain fall to the surface of our planet, supposedly at a constant rate. The concentration of these elements, therefore, should provide a measure of how long the clay blanket took to be laid down. A low concentration, for instance, would indicate a rapid accumulation of the sediments.

The results were spectacularly out of left field. Instead of measuring small amounts across the clay unit, the concentration of elements leapt to far higher levels at the K-T boundary than could be modelled by a constant sprinkling of meteoritic dust onto the Earth's surface. Iridium, for instance, was found to increase by between 40 and 330 times the background level at the different sites. Clearly some other mechanisms had to account for this extraordinary concentration.

The Alvarez team proposed that the only viable possibility was a meteorite impact. A meteorite 10 ± 4 km across would have left the amount of iridium found in the dark K-T clay layer. Such a catastrophic event would have injected about 60 times its mass into the atmosphere as pulverized rock. A proportion of this would have remained in the atmosphere for months, if not years, blocking out the Sun's rays. This would have had similar cataclysmic effects to those envisaged for the Deccan Traps. But there would have also been other associated consequences. The intense heat produced by the strike would have killed everything within the immediate 500 km of an impact. As if that wasn't enough, the intensity of the generated shock waves would have led to fires around the world. As a result, vast amounts of carbon dioxide would have been released into the air, creating highly acidic rain. Life would have been in very real danger of being snuffed out.

The conclusions of the Californian group were tremendously bold. The authors could not identify an impact site. And what was to be made of the Deccan Traps?

Over time more work was done on the Deccan Traps. Excavations have now identified dinosaur remains between the lava flows. Amidst the eruptions, the environment must have been bearable enough for life to have carried on. More recent detailed argon-argon dating has also confirmed the peak activity at Deccan was 67 million years ago; approximately 2 million years before the K-T boundary. The extinction of the dinosaurs couldn't have been as a result of the Indian volcanic eruptions.

After the Alvarez group's paper, the hunt was on to identify a realistic impact site. The hypothesis was that a meteorite approximately 10 km across had struck the Earth. An object of this size would have produced an impact crater close to 200 km in diameter. But in the early 1980s, there weren't many strong contenders. In fact, to be fair, there was nothing close. A crater of the size suggested by the Alvarez team would be the largest impact site on Earth. The two best-known impact sites from around the time of the K-T boundary were a lot smaller. The Manson Crater in Iowa in the USA was only 35 km in diameter, while the Kara Crater in the Russian Arctic was 65 km. Although smaller than predicted by the Californian group, these became the favoured candidates.

Early attempts at potassium-argon dating the Manson Crater had suggested an age of 70 million years, while the Kara Crater was supposedly 60 million. Both of these were ballpark figures and not precise enough to say whether they were coincident with the K-T boundary. By the late 1980s, argon-argon dating had been tried out on both sites. New ages of 66 million years were measured, making them both realistic contenders. Was it possible that instead of one large meteorite making a big impact crater, there could have been a storm of them striking the Earth at the same time? The result would have been several smaller impact craters on the surface.

But dating impact sites with argon-argon dating is notoriously difficult. The lingering heat of the impact speeds up the alteration of minerals, limiting the availability of grains that can be dated. As a result, samples are often found to be too altered to provide a reliable age. Dating in the 1990s showed that the original 66 million year ages for the craters were plain wrong. Redating of the Manson Crater proved it was 74 million years old. The Kara Crater was found to be 70 million years. Neither could have played a role in the K-T boundary extinction of the dinosaurs. Everything was up for grabs.

In the mid-1980s, Canadian geologist Alan Hildebrand and his advisor William Boynton at the University of Arizona, started investigating the Caribbean as a potential impact region for the K-T boundary. Their work showed that in Haiti, unlike the rest of the world, this event was half a metre thick, representing a deposit that must have been laid down near to where the impact took place. Hildebrand and Boynton argued that the source of the impact couldn't be more than 1000 km from Haiti. Hildebrand soon became interested in a geological feature called Chicxulub in Mexico.

In the 1960s, Mexico's national oil company PEMEX had been coring in the Yucatán. They found a circular feature

180 km wide and 1.5 km below the surface. At the time of discovery, it was thought to be volcanic in origin, in spite of the geology of the region. The feature was certainly the correct dimensions for the hypothesized K-T impact site. Much larger than the Manson and Kara Craters, Chicxulub was close to the 200 km diameter the Alvarez team had predicted. Gaining access to the original cores and their log reports, Hildebrand soon identified geological evidence for it being an impact site, including quartz grains that had clearly experienced tremendous pressure – 'shocked' quartz – and melted rock.

To test whether the formation of the Chicxulub Crater was coincident with the K-T boundary, its age had to be determined. Argon-argon dating was done on glassy beads found at the bottom of the crater by Carl Swisher, from the Berkeley Geochronology Group, and colleagues. Because of the earlier uncertainty with dating impact crater sites, as seen at Manson and Kara, it was crucial that the final ages were rock solid. To test the accuracy of this method, single grain samples were step heated with a laser. The argon gas was collected at increasing temperatures and measured to get a series of independent ages. Multiple ages were obtained on each sample, allowing any contamination to be easily detected and removed before the calculations were made.

The results reported in 1992 were pure dynamite. The Chicxulub Crater turned out to be 64.98 ± 0.05 million years old. This was statistically indistinguishable from the ages of 65.01 ± 0.08 and 65.07 ± 0.1 million years obtained from K-T boundary sites dated by the same method.

The dating had clinched it. Finally, the 300-year-old riddle of what had happened to the 'terrible lizards' had been solved. Ironically, they weren't wiped out by a uniformitarian mechanism but by the mother of all catastrophes – a meteorite. Science would never look at the sky in quite the same way again.

Chapter 11

TOWARDS THE LIMITS OF TIME

> *I saw Eternity the other night,*
> *Like a great ring of pure and endless light,*
> *All calm, as it was bright,*
> *And round beneath it, Time in hours, days, years,*
> *Driv'n by the spheres*
> *Like a vast shadow mov'd; in which the world*
> *And all her train were hurl'd*
> HENRY VAUGHAN (1622–1695)

The question of how old the Earth is has obsessed generations for millennia. Throughout the course of history, groups and individuals have made a grab for immortality by trying to solve the riddle of our planet's age. Just pull a number out of the air and the chances are it's been used as an age for the Earth: the ancient Hindus believed the world went through 4,320,000-year cycles of life and destruction, which would have put the Earth at 1,972,949,101 years old in AD 2000; the Persian philosopher Zoroaster believed the world to be around 12,000 years old; and the Central American Maya calculated a time of Creation equivalent to 13 August 3114 BC.

Christian culture has a particularly long history of trying to find the age of the Earth; much of it using the Bible as a source. One of the better known efforts was by Julius Africanus, who lived between AD 200 and 225. Africanus believed that all prehistory could be described as a 'cosmic' week, with each 'day' of creation lasting 1000 years. Africanus reasoned Jesus Christ had come on the sixth day, and so dated the Earth's formation to 5500 BC. The *Anglo-Saxon Chronicles* had a stab at the great question and records in AD 6 that: 'From the beginning of the world to this year, 5 thousand and 200 years had gone.' By the sixteenth century, Martin Luther had suggested the time of the Creation was 4000 BC. This view had become so widespread that even Rosalind in Shakespeare's *As You Like It* remarked 'The poor world is almost six thousand years old'.

All the attempts at using the Bible as a historical source for finding the age of the Earth used the same principle of constructing a list of Old Testament individuals and the number of years they lived. Starting with Adam, who impressively was claimed in Genesis to have had his first son Seth when he was 130 years old and to have lived to the ripe old age of 930 years, they added everyone's lifespans together. Of all the different attempts, the Protestant Bishop Ussher of Armagh is forever remembered as the individual who took this approach to its logical extreme.

Becoming head of the Anglo-Irish Church in Ireland at the age of just 25 years, Ussher was keen to demonstrate the superiority of the Protestant Christian faith over the old order. Using a library of ancient texts, many of them Greek and Roman, Ussher linked the floating chronology of biblical characters to a known point of time. The moment he hit upon was the destruction of Jerusalem by the Babylonian King Nebuchadnezzar in the sixth century BC. Ussher was an excellent historian and one of the first to realize Dennis the Little's mistake in AD 525 over the birthdate of Christ (Chapter 2). The result of all this was that Ussher could gently nudge the age of the Earth back four years on the date proposed by Martin Luther.

The date and day in the year of Creation was a little trickier. It was believed that God would have created the cosmos at a moment of symmetry between the Sun and the Earth, that is, either during a solstice or an equinox. Genesis remarks that when Adam and Eve entered the Garden of Eden, the fruit was ready to eat. Ussher took this to mean that the date of Creation must have been at the autumnal equinox in the northern hemisphere. If God had rested on the seventh day, which under Jewish tradition was a Saturday, Creation must have begun on a Sunday.

Using published astronomical tables, Ussher worked out that the autumnal equinox for the year of Creation was Tuesday October 25. This was just one day earlier than the traditional view that the Sun was formed on the fourth day – believed to be Wednesday. Close enough. Entertainingly, and much to the confusion of future historians, Ussher was notoriously suspicious of 'papists'. He completed his calculations in the Julian calendar system that was still in operation in the British Isles; hence the strange date in October for the autumnal equinox. In short, Ussher announced in AD 1654: 'Which beginning of time according to our Chronologie, fell upon the entrance of the night preceding the twenty third day of Octob. in the year of the Julian Calendar, 710.'

In the seventeenth century, Ussher and other historians used the rather abstract concept of the 'Julian period', not to be confused with the Julian calendar. This was an imaginary point in time that pre-dated the Creation. Originally, the Julian period allowed historians to compile 'dates' from different documentary sources – regardless of their religious or cultural provenance – to develop a record of the history of the Earth. Using the Julian period scheme, Ussher dated the Creation to 710 years after year zero, or, as we'd write today, 4004 BC. Although since held up to ridicule by generations, Ussher's date threw down the gauntlet to the earliest scientists.

By the eighteenth century, there were mutterings in Europe that this date could not be right. In 1721, the French Baron of Montesquieu wrote under a pseudonym in his published satire of France, the *Lettres persanes*: 'Is it possible for those that understand nature and have a reasonable idea of God to believe that matter and created things are only 6000 years old?'

By the mid-1700s, philosophers were throwing in their two-penn'orth: the Frenchman Denis Diderot suggested millions of years, while the German Immanuel Kant concurred in 1755 that the Universe must be of the order of millions of years old.

Probably one of the best-known objectors was the Frenchman Georges-Luis Leclerc, also known as the Comte de Buffon. He did experiments on the Earth's internal heat and the rate of cooling necessary to take a molten planet to reach the temperature of today. It had long been known that the deeper underground you went, the warmer it became. Buffon used this observation and his own experiments on the rate of cooling of a red-hot ball of iron to calculate an age of 75,000 years old. Uproar forced Buffon to retract his suggestion, although privately he felt this must be a minimum age. While on the rather young side compared to today's reckoning, Buffon's work was the first attempt to use scientific observations rather than 'historical' documents to date the Earth.

In 1788, James Hutton first proposed in an article that was the forerunner to his book *Theory of the Earth*: 'The result, therefore, of our present enquiry is that we find no vestige of a beginning, no prospect of an end.' As far as Hutton was concerned, for the world to be the way it was by uniformitarian principles, the timescale was so large it was impossible for him to conceive.

The problem was that by the mid-nineteenth century, Charles Darwin needed to argue for what he felt was a reasonable amount of time to allow evolution to produce today's myriad life forms. Back then, no one knew what a 'reasonable amount of time' was. In the 1859 first edition of *Origin of Species*, Darwin strayed into a bitter fight: he used the rate of erosion of the Weald in southern England as a guide. Noting that the North and South Downs once formed a continuous dome of chalk, he reckoned it had probably taken 306,662,400 years, 'or say three hundred million years' for them to reach their present form.

Within a month of the first edition's publication, he was under attack. Reviewers argued that the erosion rate Darwin suggested could have been significantly different in the past. Darwin spent much of the rest of his life agonizing over the time needed for evolution and the conflicting age estimates for the Earth. By the third edition of *Origin of Species*, reference to the Weald erosion was removed and replaced by a general statement on the enormity of time necessary for evolution.

A towering personality was soon to dominate the debate. Lord Kelvin, born as William Thomson in Belfast in 1824, was elevated to the peerage in 1892. A brilliant man, he was a world leader in virtually all fields of scientific endeavour. Physicist, engineer, Professor of Natural Philosophy at Glasgow University, Kelvin's research fuelled the construction of the first transatlantic cables. He had so many patents he died a rich man in 1907. In 1862, Kelvin turned his attention to the age of the Earth, because he was so exasperated with geologists, particularly Darwin, who he felt were ignoring the basic laws of physics. He had jointly defined the second law of thermodynamics, stating that whenever energy is converted from one form to another, a proportion is lost as heat. As far as Kelvin was concerned, physical processes on the Earth and throughout the Universe had been literally running down since their creation.

Kelvin took Buffon's original premise and assumed that the Earth had started as a molten ball and gradually cooled to its present state. Accepting that solid rock is more dense than liquid, Kelvin reasoned that solidified rock would sink from the ancient Earth's surface. The hypothesis argued that the sinking process would have created convection currents and maintained an even distribution of heat throughout the planet before the Earth eventually formed into a solid ball. All depths would therefore have the same temperature. Using the latest scientific data available on how heat migrates through rock,

Kelvin calculated how much had been lost from the Earth's surface through conduction into space. He could then calculate when the Earth was formed.

Because of some of the assumptions and uncertainties in his method, Kelvin conservatively suggested a range of ages for the Earth between 20 and 400 million years, with a calculated average of 98 million. Kelvin's conclusion: geologists should move away from the almost limitless time suggested by Hutton.

Over the next 40 years, as new data was collected on the temperature of the Earth, Kelvin repeatedly revised his age estimates, usually to the lower limits. By 1876, it was at most 76 million years, and by 1897 the age was closer to 20 million. These progressively younger 'ages' gradually reduced the number of supporters Kelvin had within the geological community. To many field geologists, the evidence could not support such young ages but they had no way of proving it.

Kelvin's ages were also a direct challenge to Charles Lyell, one of the key personalities involved in the ice age debate. Lyell was the champion of uniformitarianism, which required vast amounts of time to form the world we see today. He was inspired by the work of James Croll, in particular the importance of the Earth's changing orbit around the Sun in driving ice ages (Chapter 7). Perhaps this could form a basis for dating the Earth.

⧗

In 1867, Lyell published his tenth edition of *Principles in Geology*, arguing that the last ice age must have been between 750,000 and 800,000 years ago. On this basis, Lyell suggested that 95% of all modern seashells are found in one-million-year-old deposits. He argued that it took this length of time for a one-twentieth revolution in a species. Remember, to Lyell, cyclic changes in life were entirely consistent with uniformitarianism. So if each complete rotation of a species took 20 million years and there were 12 complete revolutions, he calculated that the explosion in life preserved at the start of the Cambrian period must have been around 240 million years ago. This was much older than what Kelvin was suggesting.

Not generally remembered for dating the age of the Earth, James Croll now waded into the debate. He saw no reason for geologists to have an 'infinite' amount of time. As far as he was concerned, the geologists' 'estimates' for rates of change were just guesses. Croll was comfortable with an upper limit of 100 million years for the age of the Earth. In contrast to Lyell, Croll assumed the last ice age happened at the most recent period of high eccentricity, which he calculated to have ended just 80,000 years ago. Lyell had ignored this, as he argued there wasn't enough time for the world to have developed as it is now if the last ice age was only 80,000 years ago. With the younger estimate, the rate of species revolution envisaged by Lyell could be reduced. The time since the start of the Cambrian period crashed down to just 60 million years. For Croll, this was far more like it.

These numbers were looked at in great detail by several prominent scientists, including Alfred Wallace, who, like Darwin, was troubled by the numbers being proposed for the age of the Earth. Allowing for the Precambrian period, when no life existed on Earth, being three times the length of the Cambrian, Wallace did his own calculations: life had existed for 24 million years and the total age of the Earth was 96 million years. Wallace felt this squared the circle. Darwin's requirement for a long lead-in time before life developed was satisfied, while Kelvin's original estimate of 98 million years was met. Darwin remained unconvinced.

Meanwhile, many geologists in Britain and America were trying a different angle. By adding up the thicknesses of all the geological units they could find and making assumptions on rates of sedimentation, they tried to get an independent age.

A whole host of numbers turned up in the literature at different times: in 1860, 96 million years for the Ganges Basin; in 1878, 200 million years for age of the Earth. None of these results seemed to make much of an impact, largely because they were known to be minimum ages and had huge uncertainties.

Back in the eighteenth century, the British astronomer Edmond Halley had questioned the age of the Earth suggested by Bishop Ussher. Halley reasoned that the observed rates of erosion must mean that the Earth was significantly older than 6000 years. He suggested a different way of calculating the age based on his observation that lakes with no rivers exiting them were very salty. The rivers entering such lakes were the most likely source of the salt. In 1715, he suggested that 'tis not improbable that the ocean it self is become salt from the same cause'. Halley reasoned that if the saltiness of the sea was measured, and it was assumed the oceans were originally freshwater, the rate of delivery of salt would provide an age for the Earth. Halley lacked the data to do the calculation.

Between 1899 and 1901, the Irish geologist John Joly at Trinity College in Dublin took Halley's idea and calculated the rate of delivery of salt to the ocean. Joly reasoned that as salt formed only a small component in rivers, he could divide the total amount in the world's seas by the rate of delivery. Joly's calculations put the age of the Earth at between 90 and 100 million years; bang on Kelvin's original suggestion.

We now know salt is massively recycled: major geological formations lock salt out of the system while vents under the sea at the edge of plate boundaries feed large amounts of salts in. Joly, one of the last main stalwarts of Kelvin's age estimate, continued reporting the results from the sea salt method and denying older ages for the Earth right up until his death 30 years later.

One of first to realize the possibilities of radioactivity for solving the age of the Earth was the New Zealander Ernest

Rutherford working in McGill University, Canada, during the early 1900s. Rutherford recognized that the enormous amount of energy contained in radioactive elements would maintain the high temperatures within the Earth. It was no longer necessary to consider the planet as the cooling body Kelvin envisaged (Rutherford later earned a Nobel Prize in 1908 for his radioactive research, ironically for chemistry, which he believed inferior to physics).

In 1904, Rutherford gave a talk to the Royal Institution in London. Who should be sitting in the audience but Kelvin. He apparently fell asleep at the start of the lecture until Rutherford came to a crucial point on the age of the Earth. Kelvin suddenly sat bolt upright, wide awake. Just then a flash of inspiration came to Rutherford. He pointed out that Kelvin had stated in earlier work that his age estimates could be incorrect if another source of energy to those known at the time of his calculations was discovered (although Kelvin had spent a large amount of effort arguing why this was unlikely). Rutherford proposed that this extra source of energy was radioactivity. Kelvin appears to have been pleased with the reverence Rutherford paid him but always maintained his age estimate was correct. He confided in a friend that it was probably the greatest contribution he had made to science.

The discovery of radioactivity led to a whole host of new elements being identified in the early twentieth century. In addition to uranium (which had been discovered in 1789), there was now radium, polonium, radon and thorium. Could these different elements be used to date the age of the Earth? In 1907, Rutherford hypothesized that helium gas was a by-product of radioactive decay, which was confirmed a year later. Assuming helium gas was trapped in the rock after being formed and its production rate could be calculated, it should be possible to work out when the rock cooled and solidified (the same principle as potassium-argon and argon-argon dating).

Rutherford tried it out. He heated a lump of a mineral

called thorianite, collected the helium gas and found the sample must have formed at least 500 million years ago, smashing Kelvin's estimate. And this wasn't even the oldest rock Rutherford could find. It was only a minimum age.

Over time, physicists worked out the series of different elements that formed when uranium decays: the 'decay chain'. Importantly, the only uranium known about then was ^{238}U and its half-life was worked out to be around 4.5 billion years. This could take science back to the dawn of time. At last, geologists had a tool to date the origins of the Earth.

⧖

Up to this point I've tried to avoid descriptions of isotopes as they would be written in scientific papers. It can get a bit heavy if you're not used to them. Unfortunately, to understand how the accepted age of the Earth was finally worked out, it's going to be necessary to cross to the other side. This final part of the story is full of different isotopes of the same elements that can all look the same after a while if you're not careful. Just keep an eye on the numbers that come in the top left-hand corners with the 'U's and the 'Pb's. I'll try and keep them to the absolute minimum.

In 1905, the American scientist Bertram Boltwood realized that lead was the end result of the uranium decay chain. He had a new idea for dating rocks. By 1907 Boltwood had got hold of 26 different rock samples to date using the uranium-lead method.

The principle of this method is that as atoms disintegrate through the decay chain, different forms of radioactive decay take place. Emissions of helium, electrons or other forms of energy take place as the atoms change from one form to another. Eventually, they reach the end of the series to form isotopes of stable lead, ^{206}Pb. By assuming no lead was present when the rock samples had crystallized, Boltwood measured the ratio of uranium to lead and showed that they had formed up to 570 million years ago. Rutherford's minimum age wasn't looking so ridiculous after all.

But it was the British geologist Arthur Holmes who took up the quest and almost single-handedly led the use of radioactive isotopes to calculate an age for the Earth. Starting in 1911, he strived to date the age of the Earth and develop a timescale for all the geological boundaries that had long been identified but remained undated. Before him, it was possible to select almost any age for the Earth. Bishop Ussher's estimate was no longer taken seriously, but Kelvin's maintained its support in some quarters. By 1931, Holmes was arguing that the age of the Earth lay somewhere between 1460 and 3000 million years.

By the late 1920s, Rutherford had shown that uranium had another isotope that had gone unrecognized: ^{235}U, which produces its own stable version of lead, ^{207}Pb. Another type of lead was also found: ^{204}Pb. But this wasn't the product of uranium decay; its concentration had not changed since the Earth had formed. No matter what uranium did, the amount of ^{204}Pb remained the same.

An important point was now realized. Because ^{235}U has a half-life of 704 million years, it decays six times faster than ^{238}U. The end result was that the older a rock sample, the higher the original ^{235}U and the more ^{207}Pb that formed. Now if we remember that the two isotopes of uranium produce different versions of lead, the ratio $^{207}Pb/^{206}Pb$ will also get bigger over time. The amount of uranium was no longer needed to get an age – only the different lead types. Well, that was the theory. It all hinged on knowing what the original mix of the different isotopes of lead had been at the time of the Earth's formation, before the decay of uranium had added to it. This was needed as the baseline. The sample had to be uranium-free.

Holmes took a sample from Greenland. This was felt to

contain lead that wasn't produced by radioactive decay and represented conditions at the time of the Earth's formation. In 1946, he reported a minimum age for the Earth of 3000 million years, which he recalculated as 3400 million years in 1947. By projecting his values back in time, he also calculated that uranium started to decay 4500 million years ago, giving a maximum age for the Earth.

After Holmes' major efforts, it was realized that no rocks on the surface of our planet, including those from Greenland, were from its year zero. On the ever-changing surface of the Earth, old rocks were constantly being destroyed and recycled. Holmes' Greenland rock gave just a minimum age. What was needed was one that represented the start of the Earth's formation but had escaped geological processes. Depressingly, it seemed that none such was likely to turn up on Earth.

During the 1940s and 50s, all parts of our solar system were thought to have formed at virtually the same time. We now know the process was considerably more complicated and took tens of millions of years, but for the timescale involved this reasoning is fine for our purposes. The idea was that because iron meteorites were the most primitive of all material in the solar system, they must have formed first. Slowly, it was argued, our planet would have coalesced through the bombardment of small, solid planetary bodies until the Earth began to resemble what we know today. Researchers reasoned that because the iron meteorites contained virtually no uranium, any lead present could not have been formed by radioactive decay. So, by analysing iron meteorites, it was possible to work out the original, primordial, lead blend of the nascent solar system, including the Earth.

By the early 1950s, American geochemist Claire Patterson, working at the California Institute of Technology, had completed the measurements on an iron meteorite and worked out its average lead composition. In 1953, he took these numbers to be the primordial mix. From this, Patterson was able to work out how much lead had been created on Earth from uranium decay and over how long. The Earth had an upper age limit of 4600 million years.

But was this the real age of the Earth? Although it seemd reasonable that meteorites were formed at the same time as our planet, this was not absolute fact in the 1950s. In 1956, Patterson strove to prove that meteorites were representative of the Earth. He measured other types of meteorite that did contain uranium. As a result of all this, Patterson could plot up the ratio of $^{207}Pb/^{204}Pb$ against $^{206}Pb/^{204}Pb$. Because the ^{204}Pb was not formed from uranium decay, its abundance remained the same. So the ratios from the different meteorites increased over time depending on how much uranium they had from the start. Together they fell on a straight line. Patterson then reasoned that the ratio of lead isotopes in the ocean floor should reflect the average makeup of the land. After all, the ocean's bed is formed from material swept into the sea by the rivers draining the eroding continents. Patterson was able to show that different samples fell on the same line as those made by the meteorites. They all had to have formed at virtually the same time. This was the clinching proof that the meteorites and the Earth had formed together.

The 4600 million year age obtained from extraterrestrial material was truly a reliable measure of the age of our planet. As the Scottish geologist James Hutton had claimed, the timescale was virtually infinite.

Epilogue

TIME'S UP FOR CREATIONISM

The clock has stopped in the dark
THOMAS STEARNS ELIOT (1888–1965)

When I started this book, I was concerned that science was not being effectively communicated. It seemed to me a real danger that society was enjoying the benefits of knowledge without understanding how it was gathered. I still feel this is a very real problem. You often hear folk bemoaning that science is 'too hard' or 'too difficult'. This is a great pity. Science is terribly exciting and I hope that by writing this book, I've given a few insights into this exhilaration. Science has a tremendous amount to offer to improve the quality of life for all of us on this piece of rock we call home. The need is an urgent one.

Our planet is now facing some of its greatest ever challenges. Recent estimates of the number of species becoming extinct are appallingly high. Somewhere between 25,000 and 50,000 species are believed to be disappearing into oblivion each year; many without even being properly identified. The numbers are so extreme it appears to be giving some of the other great extinctions we've looked at a run for their money. When you then add into the mix the prospect of catastrophic future climate change, we have some pretty taxing times ahead.

An example of where poor science understanding is exploited by vested-interest groups is with creationism, the most extreme form of which is the 'young Earthers'. Its supporters use a plethora of techniques to convince people the world is 6000 years old. They'll often be extremely selective of the studies they use to bolster their arguments and present this to people who are poorly educated in science. In fact, the whole case in support of creationism consists of muddled arguments, incomplete summaries of research, and scientific quotations given out of context. Recent discoveries in human evolution are a case in point. They give an entirely different perspective on race to that implied by creationism. As recently as 30,000 years ago, there were up to four species of human on our planet. The fact that only one now exists shows we're extremely fortunate. There is no preordained reason why we are here and the others became extinct. Denying the fossil finds and their age ignores this earlier diversity in humanity.

I experienced this first-hand when the finding of *Homo floresiensis* – the Hobbit – was first reported to the world's media at the press conference we held in Sydney in 2004. I had returned to fieldwork in northern Queensland the same evening as the press conference and over a beer had discussed the implications of the work with my colleagues in the camp-site. The following morning we found a creationist pamphlet left outside our accommodation arguing against human evolution. Clearly one of our other fellow campers had felt aggrieved at the previous night's discussion, although who carries this material on holiday is beyond me. A basic premise of the text claimed that Piltdown Man was a fraud and as a result science had failed to justify its case. I was amazed and bemused that this was seriously being used to support creationism and it is worth briefly looking at Piltdown Man and how dating showed it was a fake.

Piltdown Man refers to three sets of skeletal material found at the turn of the twentieth century by Charles Dawson, an amateur British archaeologist based in Sussex. In 1912 and in collaboration with Arthur Woodward, the Keeper of Geology at London's Natural History Museum, Dawson reported finding a cranium at the small Sussex village of Piltdown. This consisted of a human skull and ape-like jaw apparently from gravels believed to be up to two million years old. They named the find *Eoanthropus dawsoni* and it was heralded as the missing link between ape and man predicted in *Origin of Species* by Darwin. At the time, little human fossil evidence had been uncovered to support Darwin's thesis and the new find seemed to fit the bill. More excavations at the main site found other remains, with tools – including the infamous 'cricket bat' – and animal bones. Dawson later found skeletal material at a further two sites which he reported to Woodward.

Dawson died in 1916 and no more finds were made, in spite of Woodward doing 21 years more fieldwork in the area, much of it in his retirement. Over time, however, the skeletal material associated with Piltdown Man became something of an oddity. While Woodward was alive, few anthropologists were allowed to view the specimens, despite the fact that new fossil finds in other parts of Europe and Asia were in direct conflict to *Eoanthropus dawsoni*. These new finds suggested that human-like teeth and jaws were an early development in human evolution, while the braincase and forehead had apparently changed more slowly; the opposite to that seen in Piltdown Man.

After Woodward died in 1944, more stringent tests were done on the skeletal material, many of which were not available at the time of its discovery. The tests included the radio-carbon dating of different parts of the cranium. These studies soon found the Piltdown Man was a forgery, most probably perpetrated by Dawson. It consisted of a modern human skull and an orang-utan's jaw, both only several hundred years old.

⧖

Young Earth creationists believe that time, our planet and the Universe all originated from a single moment in the past. Although this view was widely held several hundred years ago, it became unsustainable when investigations of the night sky became commonplace. In 1718, Edmond Halley used observations dating back to the first century AD and realized the relative position of the stars wasn't constant over time. Importantly, he grasped that this was different to the precession of the equinoxes we looked at in Chapter 4 with dating the pyramids. Halley saw that some stars had moved relative to the others. What was going on?

Halley's ideas were developed further in the 1860s, when the British couple William and Margaret Huggins started studying the makeup of stars. By using a spectroscope, they broke up light from the star Sirius into its constituent parts so that the spectrum of colours could be seen. The Hugginses realized that, overall, the mixture was the same as that from our Sun. But in the case of Sirius, the wavelengths of the different parts had shifted to higher values: they'd moved towards the red end of the spectrum – 'redshift'.

The Doppler effect that causes redshift is identical to what happens with sound waves. Try to remember the last time you were standing on a pavement and a police car shot past you with its siren on. As the car approaches, the ear-piercing pitch increases: the wavelength gets shorter. The source is getting ever closer to your ear, so the sound waves get bunched up. But when the vehicle disappears past you, you can remove your fingers from your ears because the pitch drops off: the wavelength has got longer. In effect, the sound waves get stretched out as they reach you because the source is moving away. Fortunately, this Doppler effect can be mathematically modelled. The Hugginses were able to show that Sirius was moving away from the Earth at the speed of around 45 km per second.

During the early twentieth century, astronomers carried on making redshift measurements. So much so that by 1931, the American scientists Edwin Hubble and Milton Humason were able to prove that up to and beyond 100 million light years away, galaxies were increasingly accelerating from the Earth the further away they were. The wider implications were sensational. They suggested that if time was rolled back to its very origin, everything in the Universe must have been concentrated in just one small area of space. A history and description of this is described in Simon Singh's excellent *Big Bang*.

It's worth just looking briefly at what would have happened at around the moment of expansion, because it directly impacts on some key creationist arguments concerning the origin of time. During the 'big bang', the temperature must have reached trillions of degrees centigrade; the early Universe would have been made up of light and an almost infinite number of atomic particles. As the expansion continued, protons equivalent to the nucleus of hydrogen would have reacted with other energetic particles to form helium, and scattered off energetic electrons and light. After around 300,000 years, the temperature probably dropped to about 6000°C; low enough for the free electrons to slow down and allow light to travel without hitting anything else. Light struck upon a constant speed of 299,792 km per second and hasn't slowed down since.

Meanwhile, some areas of the universe became dense enough to attract more matter to form the first stars. As the expansion continued, stars carried on forming, living and then dying. Importantly for us, heavier elements than hydrogen and helium were produced by thermonuclear reactions during the course of a star's life and death. Almost everything we see about us is the result of a star's life cycle: the metal for our spoon at breakfast; the oxygen we breathe; the very carbon that makes us. The origins of all these elements and more are the products of an extraterrestrial process that took place before our planet was even formed. We are the consequence of at least one generation of stars that have gone before us. The Earth could not have been formed at the beginning of time.

Different methods are used to date the big bang. Many are

based on measuring the distances between different constellations as the Universe continues to expand and calculating the time required for them to have expanded from a single point in space. The most recent age estimate for our Universe was reported in 2003, with the start of time at 13.7 ± 0.2 billion years ago – all based on the background microwave fluctuations that are a hangover from the big bang; nothing to do with a guess suggested in a well-known 2005 hit single.

Even looking up at the night sky, we see time in action. Awe-inspiringly, the virtually infinite numbers of stars we see now represent light emissions that were made millions of years ago. These sparkles don't reveal what a star is like today. Imagine for a moment, an alien astronomer over 65 to 251 million light years away from the Earth looking towards our planet with a powerful telescope: the light being reflected off our surface would show that dinosaurs inhabit our planet. We all travel back in time when we look at the stars; we just don't often realize it.

Understandably, many creationists struggle with all this. They often fall back to the position of ignoring most of it and instead proposing that the speed of light has been drastically slowing down since creation. There is no evidence for this. If this were the case, it is extremely unlikely that life, or even this book, would exist. Many people are familiar with Einstein's famous equation of '$E = mc^2$' from his special theory of relativity, although perhaps not fully understanding what it represents. Einstein's great insight was that matter (m) and energy (E) are different forms of the same thing and are therefore interchangeable. To work out the amount of energy in matter, the mass has to be multiplied by the square of the speed of light (c). This last term means that just a small change in the speed of light can have a disproportionately large effect on the amount of energy produced from radioactive decay.

So to compress 13.7 billion years into 6000 years, the speed of light would have had to have been several orders of magnitude higher. This might compress the timescale but opens up a whole can of worms in other areas. For a start, a greater speed of light would also have massively increased the rate of radioactive decay, producing fatal amounts of heat on the Earth. The output of the Sun would also have been grossly increased, due to the greater rate of hydrogen fusion, producing so much extra energy that the Earth would have been incinerated. Could our ancestors have survived the onslaught of a vaporizing planet?

Ultimately, if the speed of light has changed so drastically, it would redefine the entire way time, the origin of life, the Universe and everything, is understood and taught. As Ian Plimer of the University of Melbourne states so eloquently:

> All the creation 'scientists' have to do is substantiate their claim that the speed of light has been decreasing. For this, the rewards would be instant scientific fame, universal acceptance of creation science and a Nobel prize for the creation scientist who was able to demonstrate that the cornerstone of all science was hopelessly wrong.

Needless to say, no such thing has been done.

⧗

Understanding the past gives us an opportunity to learn from yesteryear. By putting a framework of time onto past events, we can see if catastrophes hold any clues for how we should respond. By even countenancing the Earth as 6000 years old, as creationists might have us do, we risk ignoring the very lessons that may help us to successfully negotiate these future challenges.

Let's take an example of how we might learn from our ancestors by focusing on their response to relatively small climate changes in the past. This is from some work I did with colleagues at Queen's University Belfast, using the climate record from the Irish trees we looked at in Chapter 6. This impressive reconstruction of the past stretches back year by year to 7468 years and has taken some 30 years' worth of back-breaking work (not by me, I hasten to add), extracting oaks buried in bogs across Northern Ireland. Over this time, the research group at Queen's University found there were curious periods when hardly any of the trees seemed to grow at all. Other times seemed to show a tree-like Utopia where even boggy environments could be colonized: the population soared. When we looked at this more closely, we realized that there was a climate signal in this variability. During times of lots of trees, the climate was dry enough that they could move onto the bogs and flourish. When it got too wet and the water tables on the bogs rose, the trees died and no saplings could get established: the numbers crashed. Because of Ireland's position on the western seaboard of Europe, its climate is highly sensitive to what's happening in the North Atlantic. If the ocean sneezes, Ireland catches a cold. When the Atlantic was spluttering in the past, the trees seemed to show the land was verging on pneumonia. But if there were times in the past when the trees weren't happy, how did the Irish people feel?

Looking at the archaeological record, we were fortunate that there was over 50 years' worth of radiocarbon dates reported from excavations. We had access to over 450 measurements completed on forts, crannogs (homes built on artificial islands on lakes and in marshes) and settlements. We converted the ages to calendar dates so we could directly compare when these structures were being built to the climate signal preserved by the trees. What we found stunned us. Refuges were constructed when the climate took a dive. Almost without fail, when times got bad, people would congregate together in permanent locations, defending what little food and other resources they had. Scenarios for climate change suggest a potentially worse future than that with which our ancestors had to contend. Can we respond more sensibly than waging war on our neighbours to steal what little they have? I hope so.

We urgently need to investigate past human responses in other parts of the world to see if the same pattern holds in different climatic areas. But if the Earth is only 6000 years old, many of these events and others we have looked at in this book never took place. We can't use these past scenarios to understand and plan for the future. I doubt many of us find this acceptable.

Until creationism provides compelling evidence for its arguments rather than blithely discounting hundreds of years of scientific research, it will continue to be a belief and should be treated as such. To allow the distortion of time runs the risk of returning to a period where indoctrination becomes the accepted norm. We owe it to ourselves and future generations to vigorously challenge so-called creation 'science'.

The past is the key to the future and we need all the time we can get to see it.

FURTHER READING

A vast amount of research has been done on dating the past. An exhaustive overview would fill several books and still not do justice to the topic. As a result I have had to be judicious in my selection of examples and sources. Below are some of the key texts under the relevant chapter headings for those readers who want to follow up some of the aspects covered in this book. Wherever possible, I have chosen excellent detailed overviews and recent articles that give a good summary of the earlier literature. In some cases, no widely accessible book is available, so I have had to resort to listing scientific articles only. Hopefully these should provide a good platform for finding other sources.

1. The ever-changing calendar

Duncan, D.E. (1999) *The Calendar*. Fourth Estate, London.

McCready, S. (ed.) (2001) *The Discovery of Time*. Sourcebooks, Naperville, Illinois.

Waugh, A. (1999) *Time*. Headline Book Publishing, London.

2. A hero in a dark age

Alcock, L. (1973) *Arthur's Britain*. Pelican, England.

Bede (1990) *Ecclesiastical History of the English People* (eds L. Sherley-Price and R.E. Latham). Penguin Books, London.

Geoffrey of Monmouth (1966) *The History of the Kings of Britain* (ed. L. Thorpe). Penguin Books, London.

Gildas (1978) *The Ruin of Britain and Other Works* (ed. M. Winterbottom). Phillimore and Co., London.

Malory, Sir Thomas (1998) *Le Morte D'Arthur* (ed. H. Cooper). Oxford University Press, Oxford.

Phillips. G. and Keatman, M. (1992) *King Arthur: The True Story*. Arrow, London.

Swanton, M. (ed.) (2000) *The Anglo-Saxon Chronicle*. Phoenix Press, London.

3. The forged cloth of Turin

Arnold, J.R. and Libby, W.F. (1949) Age determinations by radiocarbon content: Checks with samples of known age. *Science*, 110, 678–80.

Damon, P.E., Donahue, D.J., Gore, B.H., Hathaway, A.L., Jull, A.J.T., Linick, T.W., Sercel, P.J., Toolin, L.J., Bronk, C.R., Hall, E.T., Hedges, R.E.M., Housley, R., Law, I.A., Perry, C., Bonani, G., Trumbore, S., Woelfi, W., Ambers, J.C., Bowman, S.G.E., Leese, M.N. and Tite, M.S. (1989) Radiocarbon dating of the Shroud of Turin. *Nature*, 337, 611–15.

Gove, H.E. (1990) Dating the Turin Shroud – An assessment. *Radiocarbon*, 32, 87–92.

Hedges, R.E.M. (1989) Shroud irradiated with neutrons? Reply. *Nature*, 337, 594.

Libby, W.F., Anderson, E.C. and Arnold, J.R. (1949) Age determination by radiocarbon content: world-wide assay of natural radiocarbon. *Science*, 109, 227–8.

Phillips, T.J. (1989) Shroud irradiated with neutrons? *Nature*, 337, 594.

Reimer, P.J., Baillie, M.G.L., Bard, E., Bayliss, A., Beck, J.W., Bertrand, C.J.H. et al. (2004) IntCal04 terrestrial radiocarbon age calibration, 0-26 cal kyr BP. *Radiocarbon*, 46, 1029–58.

Rogers, R.N. (2005) Studies on the radiocarbon sample from the Shroud of Turin. *Thermochimica Acta*, 425, 189–94.

4. The pyramids and the bear's groin

Shaw, I. (2000) *The Oxford History of Ancient Egypt*. Oxford University Press, Oxford.

Spence, K. (2000) Ancient Egyptian chronology and the astronomical orientation of pyramids. *Nature*, 408, 320–4.

5. The volcano that shook Europe

Baillie, M.G.L. and Munro, M.A.R. (1988) Irish tree rings, Santorini and volcanic dust veils. *Nature*, 332, 344–6.

Downey, W.S. and Tarling, D.H. (1984) Archaeomagnetic dating of Santorini volcanic eruptions and fired destructive levels of late Minoan civilization. *Nature*, 309, 519–23.

Hammer, C.U., Clausen, H.B., Friedrich, W.L. and Tauber, H. (1987) The Minoan eruption of Santorini in Greece dated to 1645 BC? *Nature*, 328, 517–19.

Hammer, C.U., Kurat, G., Hoppe, P., Grum, W. and Clausen, H.B. (2003) Thera eruption date 1645 BC confirmed by new ice core data?,

Proceedings of SCIEM2000 (Synchronisation in the Eastern Mediterranean in the 2nd Millenium BC).

LaMarche, V.C. and Hirschboeck, K.K. (1984) Frost rings in trees as records of major volcanic activities. *Nature*, **307**, 121–6.

Manning, S.W. (1999) *A Test of Time: The Volcano of Thera and the Chronology and History of the Aegean and East Mediterranean in the Mid-second Millennium BC*. Oxbow Books, Oxford.

Manning, S.W., Kromer, B., Kuniholm, P.I. and Newton, M.W. (2001) Anatolian tree rings and a new chronology for the east Mediterranean Bronze-Iron ages. *Science*, **294**, 2532–5.

Marinatos, S. (1939) The volcanic destruction of Minoan Crete. *Antiquity*, **13**, 425–39.

Montelius, O. (1885) *Dating the Bronze Age with Special Reference to Scandinavia*. K. Vitterhets Historie och Antikvitetsakademien.

Pearce, N.J.G., Westgate, J.A., Preece, S.J., Eastwood, W.J. and Perkins. W.T. (2004) Identification of Aniakchak (Alaska) tephra in Greenland ice core challenges the 1645 BC date for Minoan eruption of Santorini. *Geochemistry, Geophysics, Geosystems*, **5**, DOI 10.1029/2003GC000672.

6. The Mandate from Heaven

Baillie, M. (2000) *Exodus to Arthur*. Batsford, London.

Baillie, M.G.L. (1995) *A Slice Through Time: Dendrochronology and Precision Dating*. Routledge, London.

McCafferty, P. and Baillie, M. (2005) *The Celtic Gods: Comets in Irish Mythology*. Tempus Publishing, Stroud.

Rigby, E., Symonds, M. and Ward-Thompson, D. (2004) A comet impact in AD 536? *Astronomy & Geophysics*, **45**, 1.1–1.4.

7. The coming of the ice

Berger, A. and Loutre, M.F. (1991) Insolation values for the climate of the last 10 million years. *Quaternary Science Reviews*, **10**, 297–318.

Blunier, T. and Brook, E.J. (2001) Timing of millennial-scale climate change in Antarctica and Greenland during the last glacial period. *Science*, **291**, 109–12.

Dansgaard, W., Johnsen, S.J., Clausen, H.B., Dahl-Jensen, D., Gundestrup, N.S., Hammer, C.U., Hvidberg, C.S., Steffensen, J.P., Sveinbjörnsdottir, A.E., Jouzel, J. and Bond, G. (1993) Evidence for general instability of past climate from a 250-kyr ice-core record. *Nature*, **364**, 218–20.

EPICA Community Members (2004) Eight glacial cycles from an Antarctic ice core. *Nature*, **429**, 623–8.

Gribbin, J. and Gribbin, M. (2001) *Ice Age*. Allen Lane, Penguin Press, London.

Imbrie, J. and Imbrie, K.P. (1979) *Ice Ages: Solving the Mystery*. Macmillan, London.

Imbrie, J. Shackleton, N.J., Pisias, N.G., Morley, J.J., Prell, W.L., Martinson, D.G., Hayes, J.D., MacIntyre, A. and Mix, A.C. (1984) The orbital theory of Pleistocene climate: support from a revised chronology of the marine $\delta^{18}O$ record. In *Milankovitch and Climate*, Part 1, ed. by A. Berger, Reidel, Hingham, Massachusetts, 269–305.

North Greenland Ice Core Project Members (2004) High-resolution record of Northern Hemisphere climate extending into the last interglacial period. *Nature*, **431**, 147–51.

Rohling, E.J. and Pälike, H. (2005) Centennial-scale climate cooling with a sudden cold event around 8,200 years ago. *Nature*, **434**, 975–9.

Walker, M. (2005) *Quaternary Dating Methods*. John Wiley & Sons, Chichester.

8. The lost worlds

Anderson, A. (2000) Differential reliability of ^{14}C AMS ages of *Rattus exulans* bone gelatin in south Pacific prehistory. *Journal of the Royal Society of New Zealand*, **30**, 243–61.

Elias, S.A. (1999) Quaternary biology update, debate continues over the cause of Pleistocene megafauna extinction. *Quaternary Times*, June, 11.

Fiedel, S. and Haynes, G. (2004) A premature burial: Comments on Grayson and Meltzer's 'Requiem for overkill'. *Journal of Archaeological Science*, **31**, 121–31.

Flannery, T. (1997) *The Future Eaters*. Reed New Holland, Sydney.

Flannery, T. (2002) *The Eternal Frontier*, Vintage, London.

Guthrie, R.D. (2004) Radiocarbon evidence of mid-Holocene mammoths stranded on an Alaskan Bering Sea island. *Nature*, **429**, 746–9.

Higham, T., Anderson, A. and Jacomb, C. (1999) Dating the first New Zealanders: The chronology of Wairau Bar. *Antiquity*, **73**, 420–7.

Holdaway, R.N. (1996) Arrival of rats in New Zealand. *Nature*, **384**, 225–6.

Holdaway, R.N. and Jacomb, C. (2000) Rapid extinction of the moas

(*Aves: Dinornithiformes*): Model, test, and implications. *Science*, **287**, 2250–4.

Jones, R. (1998) Dating the human colonization of Australia: radiocarbon and luminescence revolutions. *Proceedings of the British Academy*, **99**, 37–65.

Johnson, C.N. (2002) Determinants of loss of mammal species during the Late Quaternary 'megafauna' extinctions: life history and ecology, but not body size. *Proceedings of the Royal Society of London B*, **269**, 2221–8.

Jull, A.J.T., Iturralde-Vinent, M., O'Malley, J.M., MacPhee, R.D.E., McDonald, H.G., Martin, P.S., Moody, J. and Rincon. A. (2004) Radiocarbon dating of extinct fauna in the Americas recovered from tar pits. *Nuclear Instruments and Methods in Physics Research*, B223–4, 668–71.

Miller, G.H., Fogel, M.L., Magee, J.W., Gagan, M.K., Clarke, S.J. and Johnson, B.J. (2005) Ecosystem collapse in Pleistocene Australia and a human role in megafaunal extinction. *Science*, **309**, 287–90.

Miller, G.H., Magee, J.W., Johnson, B.J., Fogel, M.L., Spooner, N.A., McCulloch, M.T. and Ayliffe, L.K. (1999) Pleistocene extinction of *Genyornis newtoni*: human impact on Australian megafauna. *Science*, **283**, 205–8.

Roberts, R.G., Jones, R. and Smith, M.A. (1990) Thermoluminescence dating of a 50,000-year-old human occupation site in northern Australia. *Nature*, **345**, 153–6.

Roberts, R.G., Flannery, T.F., Ayliffe, L.K., Yoshida, H., Olley, J.M., Prideaux, G.J., Laslett, G.M., Baynes, A., Smith, M.A., Jones, R. and Smith, B.L. (2001) New ages for the last Australian megafauna: continent-wide extinction about 46000 years ago. *Science*, **292**, 1888–92.

Turney, C.S.M., Bird, M.I., Fifield, L.K., Roberts, R.G., Smith, M.A., Dortch, C.E., Grün, R., Lawson, E., Ayliffe, L.K., Miller, G.H., Dortch, J. and Cresswell, R.G. (2001) Early human occupation at Devil's Lair, southwestern Australia 50,000 years ago. *Quaternary Research*, **55**, 3–13.

Turney, C.S.M., Kershaw, A.P., Moss, P., Bird, M.I., Fifield, L.K., Cresswell, R.G., Santos, G.M., di Tada, M.L., Hausladen, P.A. and Zhou, Y. (2001) Redating the onset of burning at Lynch's Crater (North Queensland): Implications for human settlement in Australia. *Journal of Quaternary Science*, **16**, 767–71.

Willerslev, E., Hansen, A.J., Binladen, J., Brand, T.B., Gilbert, M.T.P., Shapiro, B., Bunce, M., Wiuf, C., Gilichinsky, D.A. and Cooper, A. (2003) Diverse plant and animal genetic records from Holocene and Pleistocene sediments. *Science*, **300**, 791–5.

9. And then there was one

Falguères, C., Bahain, J.J., Yokoyama, Y., Arsuaga, J.L., de Castro, J.M.B., Carbonell, E., Bischoff, J.L. and Dolo, J.M. (1999) Earliest humans in Europe: the age of TD6 Gran Dolina, Atapuerca, Spain. *Journal of Human Evolution*, **37**, 343–52.

Forth, G. (2005) Hominoids, hairy hominoids and the science of humanity. *Anthropology Today*, **21**, 13–17.

Grün, R. and Stringer, C.B. (1991) Electron spin resonance dating and the evolution of modern humans. *Archaeometry*, **33**, 153–99.

Higham, T., Bronk Ramsey, C., Karavanic, I., Smith, F.H. and Trinkaus, E. (2006) Revised direct radiocarbon dating of the Vindija G1 Upper Paleolithic Neandertals. *Proceedings of the National Academy of Sciences*, **103**, 553–7.

Huffman, O.F., Zaim, Y., Kappelman, J., Ruez Jr, D.R., de Vos, J., Rizal, Y., Aziz, F. and Hertler, C. (2006) Relocation of the 1936 Mojokerto skull discovery site near Perning, East Java. *Journal of Human Evolution* (in press).

McDougall, I., Brown, F.H. and Fleagle, J.G. (2005) Stratigraphic placement and age of modern humans from Kibish, Ethiopia. *Nature*, **433**, 733–6.

Mellars, P. (2004) Neanderthals and the modern human colonization of Europe. *Nature*, **432**, 461–5.

Morwood, M.J., Soejono, R.P., Roberts, R.G., Sutikna, T., Turney, C.S.M., Westaway, K.E., Rink, W.J., Zhao, J.-X., van den Bergh, G.D., Due, R.A., Hobbs, D.R., Moore, M.W., Bird, M.I. and Fifield, L.K. (2004) Archaeology and age of *Homo floresiensis*, a new hominin from Flores in eastern Indonesia. *Nature*, **431**, 1087–91.

Morwood, M.J., O'Sullivan, P.B., Aziz, F. and Raza, A. (1998) Fission-track ages of stone tools and fossils on the east Indonesian island of Flores. *Nature*, **392**, 173–6.

Morwood, M.J., O'Sullivan, P.O., Susanto, E.E. and Aziz, F. (2003) Revised age for Mojokerto 1, an early *Homo erectus* cranium from East Java, Indonesia. *Australian Archaeology*, **57**, 1–4.

Shipman, P. (2001) *The Man Who Found the Missing Link: The Extraordinary Life of Eugène Dubois*. Simon & Schuster, New York.

Stringer, C. (2002) Modern human origins: Progress and prospects. *Philosophical Transactions of the Royal Society of London*, **B357**, 563–79.

Stringer, C. and Andrews, P. (2005) *The Complete World of Human Evolution*. Thames & Hudson, London.

Swisher III, C.C., Curtis, G.H., Jacob, T., Getty, A.G., Suprijo, A. and

Widiasmoro (1994) Age of the earliest known hominids in Java, Indonesia. *Science*, **263**, 1118–21.

Swisher III, C.C., Rink, W.J., Antón, S.C., Schwarcz, H.P., Curtis, G.H., Suprijo, A. and Widiasmoro (1996) Latest *Homo erectus* of Java: potential contemporaneity with *Homo sapiens* in southeast Asia. *Science*, **274**, 1870–4.

Trinkaus, E., Moldovan, O., Milota, S., Bilgâr, A., Sarcina, L., Athreya, S. et al. (2003) An early modern human from the Pestera cu Oase, Romania. *Proceedings of the National Academy of Sciences*, **100**, 11231–6.

Vekua, A., Lordkipanidze, D., Rightmire, G.P., Agusti, J., Ferring, R., Maisuradze, G., Mouskhelishvili, A., Nioradze, M., de Leon, M.P., Tappen, M., Tvalchrelidze, M. and Zollikofer, C. (2002) A new skull of early *Homo* from Dmanisi, Georgia. *Science*, **297**, 85–9.

10. The hole in the ground

Alvarez, L.W., Alvarez, W., Asaro, F. and Michel, H.V. (1980) Extraterrestrial cause for the Cretaceous-Tertiary extinction. *Science*, **208**, 1095–108.

Amthor, J.E., Grotzinger, J.P., Schröder, S., Bowring, S.A., Ramezani, J., Martin, M.W. and Matter, A. (2003) Extinction of *Cloudina* and *Namacalathus* at the Precambrian-Cambrian boundary in Oman. *Geology*, **31**, 431–4.

Burnie, D. (2004) *The Concise Dinosaur Encyclopedia*. Kingfisher, London.

Cadbury, S. (2000) *Terrible Lizard*. Owl Books, New York.

Chen, P.J., Dong, Z.-M. and Zhen, S.-N. (1998) An exceptionally well-preserved theropod dinosaur from the Yixian Formation of China. *Nature*, **391**, 147–52.

Frankel, C. (2000) *The End of the Dinosaurs*. Cambridge University Press, Cambridge.

Hildebrand, A.R. and Boynton, W.V. (1990) Proximal Cretaceous–Tertiary boundary impact deposits in the Caribbean. *Science*, **248**, 843–7.

Swisher III, C.C., Grajales-Nishimura, J.M., Montanari, A., Margolis, S.V., Claeys, P., Alvarez, W., Renne, P., Cedillo-Pardo, E., Maurrasse, F.J.-M.R., Curtis, G.H., Smit, J. and McWilliams, M.O. (1992) Coeval ^{40}Ar/^{39}Ar ages of 65.0 million years ago from Chicxulub Crater melt rock and Cretaceous–Tertiary boundary tektites. *Science*, **257**, 954–8.

Venkatesan, T.R., Pande, K. and Gopalan, K. (1993) Did Deccan volcanism pre-date the Cretaceous/Tertiary transition? *Earth and Planetary Science Letters*, **119**, 181–9.

11. Towards the limits of time

Burchfield, J.D. (1990) *Lord Kelvin and the Age of the Earth*. University of Chicago Press, Chicago.

Dalrymple, G.B. (1991) *The Age of the Earth*. Stanford University Press, Standford, California.

Darwin, C. (1859) *On the Origin of Species by Means of Natural Selection*. Reprinted 1985, Penguin Classics, London.

Holmes, A. (1965) *Principles of Physical Geology*. Thomas Nelson & Sons, London.

Lewis, C. (2000) *The Dating Game*. Cambridge University Press, Cambridge.

Patterson, C. (1956) Age of meteorites and the Earth. *Geochimica et Cosmochimica Acta*, **10**, 230–7.

Epilogue: Time's up for creationism

Dawson, C. and Woodward, A. (1913) On the discovery of a palaeolithic human skull. *Quaterly Journal of the Geological Society of London*, **69**, 117–51.

de Vries, H. and Oakley, K.P. (1959) Radiocarbon dating of the Piltdown skull and jaw. *Nature*, **184**, 224–6.

Plimer, I. (1994) *Telling Lies for God*. Random House, Sydney.

Russell, M. (2003) *Piltdown Man: The Secret Life of Charles Dawson and the World's Greatest Archaeological Hoax*. Tempus Publishing, Stroud.

Singh, S. (2004) *Big Bang*. Fourth Estate, London.

Singh, S. (2005) Katie Melua's bad science. *Guardian*, Friday 30 September.

Spergel, D.N., Verde, L., Peiris, H.V., Komatsu, E., Nolta, M.R., Bennett, C.L., Halpern, M., Hinshaw, G., Jarosik, N., Kogut, A., Limon, M., Meyer, S.S., Page, L., Tucker, G.S., Weiland, J.L., Wollack, E. and Wright, E.L. (2003) First-year Wilkinson Microwave Anisotropy Probe (WMAP) observations: Determination of cosmological parameters. *Astrophysical Journal Supplement Series*, **148**, 175–94.

Turney, C.S.M., Baillie, M., Palmer, J. and Brown, D. (2006) Holocene climatic change and past Irish societal response. *Journal of Archaeological Science*, **33**, 34–8.

Weiner, J.S., Oakley, K.P. and Le Gros Clark, W.E. (1953) The solution of the Piltdown problem. *The Bulletin of the British Museum (Natural History)*, **2**, 141–6.

Woodward, A.S. (1917) Fourth note on the Piltdown Gravel, with evidence of a second skull of *Eoanthropus dawsoni*. *Quaterly Journal of the Geological Society of London*, **73**, 1–10.

INDEX

 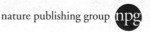

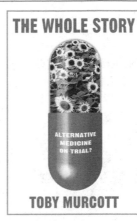
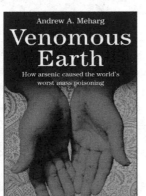

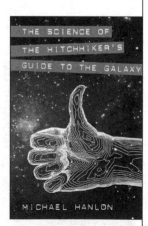